SELF-DIRECTED PROBLEM SOLVING

Idea Production in Mathematics

M. Ann Dirkes
Indiana University–Purdue
Fort Wayne

UNIVERSITY
PRESS OF
AMERICA

Lanham

University Press of America®, Inc.
4720 Boston Way
Lanham, Maryland 20706

3 Henrietta Street
London WC2E 8LU England

Printed in the United States of America
British Cataloging in Publication Information Available

Library of Congress Cataloging-in-Publication Data

Dirkes, M. Ann.
Self-directed problem solving / M. Ann Dirkes.
p. cm.
Includes index.
1. Thought and thinking—Study and thinking. 2. Problem solving.
I. Title.
LB1590.3.D59 1993 370.15'2—dc20 93–3836 CIP

ISBN 0–8191–9130–2 (pbk. : alk. paper)

The paper used in this publication meets the minimum requirements of American National Standard for Information Sciences—Permanence of Paper for Printed Library Materials, ANSI Z39.48–1984.

Contents

Preface

Long ago, I decided that I had to find a way to nurture curiosity and intellectual satisfaction. As a teacher, my task was clear. I had to get students to produce ideas for themselves so that they would activate their own brain power. The premise dealt with student idea production as a means for creating concepts and with student control for stimulating curiosity. The challenge persisted in my work with students in elementary through graduate school. ***Self-Directed Problem Solving*** responds to this challenge.

The book is designed to help students help themselves learn and invent ways to use what they know. The first step is for teachers to recognize opportunities for student self-direction and for freeing individuals from defeating habits of mind. Teachers interested in new techniques for teaching mathematics include college instructors of methods courses and many others who teach mathematics or other courses that require mathematical thinking. Those interested in psychology will find learning described as a problem solving activity. The best scenario adds teachers of elementary and secondary schools who would develop curiosity and idea production from the first encounters students have with mathematics in school.

This book is not an algorithmic approach to problem solving in computation and word problems. Rather, the strategies introduced ask that students aggressively pursue what they already know and use it to construct new concepts. College students tell me that they do not figure out how the process works for nearly half the course. When they have conquered their initial fears about new kinds of thinking, however, they find a new sense of learning and great satisfaction from their own efforts to understand mathematics. Learning to confront ambiguity is a by-product, an accomplishment that rarely occurs without anxiety.

When teachers decide to develop self-directed problem solving as a foundation for learning, they test their own propensity for risk and initiative. To help them, the book shows (a) how idea production contributes to constructive thinking, (b) how to develop self-directed thinking, and (c) how to make idea production an integral part of mathematics. Most important, it raises expectations for the thinking that students initiate and transforms the assessment of mathematics performance.

Instructors can use the entire program presented here or use selected components of it. The choice depends on the beliefs that teachers have about the nature of mathematics and its use in today's world. Those who put their faith in the hidden potential of students foster self-direction as a catalyst for problem solving they considered impossible in students learning through conventional methods.

Strategies described in the book have been effective in elementary school, secondary school and college. Students performing at a wide range of achievement levels and degrees of academic support from home learn to direct their own thinking and build upon the knowledge and strategies at their disposal. Although greater access to knowledge and formal reasoning is more likely in older students than younger ones, productive thinking and the dispositions for learning are remarkably similar. Case studies suggest that serious self-direction in elementary school will "pay off" as students go through school and enrich the thinking capacity of college students who have never learned to monitor and control this resource.

Chapter 1 describes the nature of learning and problem solving and Chapter 2 the components of the program, *Self-Direction in Mathematics (SDM)*. Components include: mathematics recommended for contemporary schools, the meaning of self-direction, idea listing, procedures for problem solving with idea production, materials, and assessment.

In Chapters 3 through 6, *SDM* strategies engage students in a problem-solving mode of learning and thinking. A comparison of routine thinking and a fluent generation of ideas centers on the purpose and effects of each. Students examine their beliefs about the kind of thinking that influences their work and their interpretations of mathematical situations.

When students plan, monitor, and evaluate to reach new levels of understanding, they use mathematical power (Chapter 3). Confronting inertia and other threats to maximum learning, levels of self-direction correlate with levels of understanding. Under self-direction, students read, draw and produce ideas to grapple with language and context. They also make decisions about their own strategies and dispositions. Motivation is power insofar as it affects the degree of challenge students accept, their curiosity, and the control they exercise over their own thinking.

Idea listing is the strategy developed in Chapter 4. It addresses two related concerns that are essential to problem solving, namely, student-initiated searches for connections and the confidence that prolongs thinking. The goal is for students to list ideas freely whenever they deal with ambiguity. Productive thinking helps them search for clarity and for new ideas that solve novel and complex problems. Rather than say, "I don't know what to do," students list ideas in many categories and make connections. Other strategies include creating a context, reading to learn, drawing to illustrate operations with fractions, marking drawings with numbers, and *listening* to fractions. A procedure for scoring lists contributes to an evaluation of thinking power.

Students use a checklist to guide their thinking into mathematical concerns, and cooperative group sessions stimulate fluency and flexibility in lists. The outcome is active participation and increased time for thinking. Students testify that productive thinking offers them a way to begin to solve problems, invent ideas, correct errors, and visualize contexts. Because lists are a record of the knowledge accessible to students, teachers use them in diagnosis and prescription.

Chapter 5 offers a flexible plan for giving structure to word problems: (a) a statement of facts, (b) restatements of the question to clarify the intent, (c) abundant idea production for possible approaches to a solution, and (d) a check on the agreement among these parts. Self-direction governs choices for quick algorithmic applications or for more extensive idea listing in each step of the plan. Drawings provide a format for thinking beyond habitual perceptions of mathematical concepts and word problems.

The problem-solving plan is expanded in Chapter 6. Open problems require the addition of facts and the statement of a question, and problems with many answers call for idea listing with original thinking and corresponding dispositions. Using the same general plan for social problems, students use productive thinking to influence out-of-school behavior. With experience, they transfer the practical thinking used outside of school to academic tasks.

The connection between assessment and learning is the focus of Chapter 7. Activities described in earlier chapters give both students and teachers information about learning power and constructive thinking in mathematics. Although tests with one-answer questions are easier to score, they are less valid for measuring the generative ability required for learning and problem solving or for supporting student communication, reasoning, and connection-making.

The book ends with conjectures for future learning and accomplishment. The self-directed problem solving students use in school mathematics prepares them to use the same strategies outside of school in new contexts and working relationships. Transferring mathematical power to students is apt to correlate with the shift of power necessary for the restructure of learning habits in a changing world.

The many students who contributed to this volume by sharing their own thinking worked through changes in learning habits. Many said that they changed the ways in which they think about mathematics and about their own mathematical abilities. They influenced one another and their instructors in college and pre-college classrooms. Most important, they helped themselves and their teachers understand how learning takes place and how strategies vary for individuals. The author is indebted to them for the courage and persistence they showed during each transformation to a higher level of self-direction.

It is instructive also to observe students who found self-directed learning so foreign to their experience that they resisted it strongly. These students may enjoy constructive thinking in large measure only if and when they experience employment or home responsibilities that demand it. Some, however, might find that life does not offer rich opportunities to stretch their abilities or that they choose not to take advantage of new challenges. A discussion of future possibilities suggests that the beliefs of both teachers and students are important because they influence classroom goals.

In this book, the program describing self-directed mathematics is designed to enrich thinking enough to have students find out what they can do. In a nation that depends on the participation of its people, good mathematics is a never ending search for new ideas and supporting evidence. Self-directed individuals who can produce abundant alternatives and improvements are equipped to contribute to society as well as initiate a response to the ambiguity and uncertainty they find in life.

1 The Problem

I don't know how to do this.

What 'do' you know about it?
I added numbers on the last problem and it was wrong.

What do you know about this problem?
There's some numbers and miles.

What else?
Fifty miles an hour?

At the moment of confusion, students decide whether to ask someone else for help or direct their own thinking in search of new understanding. How students define mathematics and what they expect of themselves suggest a course of action. Influenced by a belief that they *are not good at math* or that *they know math but do not perform well*, many students catch bits of information and rely upon teachers to put ideas together. In response, teachers decide whether to cultivate self-help or state rules and answers.

Too many students have confidence in erroneous beliefs and lose confidence in mathematics as they progress through school.[1] In response, teachers examine the nature of mathematics, its uses, and the influence of culture on student attitudes. With only modest benefits from external research, they base decisions largely on their own beliefs and conjectures.

Many decisions stem from a teacher's perception of the purpose for a mathematics curriculum. Why do students study mathematics? Are they preparing to take tests, manage skills, earn a living, or acquire strategies that last for a life time? Teachers who build programs around the power that students have over their own thinking emphasize self-direction and idea production. The advantages of independent thinking preempt the search for ideas determined by someone else. An inner drive constructs mathematics so that it has meaning for individuals.

RECOMMENDATIONS

Goals for mathematical performance are shifting from a narrow focus on routine skills to the development of broad-based power.[2] Accordingly, educators recommend that students be actively involved in doing mathematics through problem solving, communication, connection-making, and reasoning.[3] These activities include the production of ideas in conjectures and relationships, thereby changing the major focus of instruction from completed products to the ways in which students think. The scope of thinking encompasses physical and social contexts as well as personal beliefs and understanding about cognition .[4]

To sample recommendations from practitioners, fourteen teachers of grades three through five were asked to describe desirable mathematical thinking for students in today's schools. Their ideas are presented in three lists. List A includes generalizations about thinking and the recall of algorithms. The governing principle for students is to follow a plan designed by someone else.

List A: Algorithms

Systematically work toward an answer.
Work through a problem.
Put down the steps logically.
Use a prescribed formula of steps.
Recall how to solve the problem.
Use mathematics.
Decide whether the process is correct.

List B suggests a foundation for the development of mathematics in later years. There is an emphasis on understanding.

List B: A Foundation

Use prior knowledge to understand new ideas.
Understanding is the goal.
Ask, "Does the answer make sense?"
Visualize problems in people and things.
Take time to think.
Interpret numbers.
See it in more than one way.

Accept intellectual risk.

In List C, there is a sense that students use productive thinking to construct ideas.

List C: Productive Thinking

Make connections.
Expand on ideas.
Give your own ideas, draw, and explain.
Write possible solutions. Try things out.
Discover it for yourself.
Find the missing pieces.
Draw conclusions.
List what you know and organize these ideas.
Question, plan, act, evaluate.

Teachers concluded that problem solving is what students do to understand, conclude, and invent. In their view, immediate recall is only a small part of the challenge of problem solving. The rest of the process consists of producing new connections and guesses as well as other constructive thinking.

In spite of these beliefs, teachers noted that the instruction they offered students reinforced the generalizations described in List A. Only occasionally did they ask students to use the thinking suggested in Lists B and C. Some teachers said that they did not know how to make students better thinkers or that too many factors worked against it. The observations of these teachers support other accounts of practice in today's classrooms where thinking and problem solving are simplified greatly. Regardless of the mathematical challenge, both high and low achievers have depended largely on quick recall as a routine strategy. They rarely ask questions, elaborate on ideas, or use spatial representations.[5] The most common response to a challenge is a brief answer or the abandonment of problems. Some students do not believe in planning or that their efforts will make a difference. Lacking task involvement or a concern for competition, many students do not care to expand or stretch their thinking about mathematics. Low scores on tests that measure problem-solving tasks[6] underscore this claim.

When learning becomes a matter of following directions, students reduce their choices to ideas from someone else or no ideas. "Either I know it or I don't." A listening directive carries a similar message. Told to pay attention in class, students infer that they are to look at the teacher. They might be asked also to remember what they hear, but rarely are they told to make new connections. In the usual scheme of learning, a failure to ask for original thinking reinforces the failure of students to explore the abundance of their own idea production. The outcome is meager expectations for autonomous problem solving.

For many years, I have struggled with the question of designing instruction for both the development of mathematics and the dispositions of students. How much attention should each receive? Gradually my concern has shifted to the persons who learn and who choose the degree to which they influence their own learning and problem solving. The more I work with students the more I ask them to take a larger share in the direction of their own learning. Because students govern the internal drive and do the thinking that makes them learners and problem solvers, their power usurps the very best mathematical heuristics and sequence of subject matter. Perhaps my perceptions will change, but at this time I see too many examples to ignore. Students continue to teach me about their unique individual power. One theme appears again and again: These students don't know what they can do. Only their own direction will show them.

HOW STUDENTS THINK

Asked about student thinking, one fifth grade teacher remarked, "I don't believe that students think." She joins many educators who conclude that an emphasis on memory and speed inhibits mathematical understanding[7] and suppresses the informal knowledge of students.[8] When immediacy and closure dominate the curriculum, routine thinking curtails ideas original to the problem solver. Rather than try to change this pattern, most teachers limit instruction to mathematical definitions and algorithms. The decision is reinforced by students and parents who equate teaching with lecture and drill. In some schools pressures are so strong[9] that teachers feel powerless to develop the curiosity and sense-making required for significant learning.

College courses follow a similar, predictable format. Without it, many students resort to helplessness. *I don't understand* comes too quickly — even before sustained efforts to understand. *What shall I do?* preempts thinking even when individuals can answer the question themselves. Satisfying an instructor is more important, or at least more captivating, than learning.

Students preparing to become teachers were asked to make decisions about the work they would prescribe for themselves to enrich their understanding of teaching and learning mathematics. One student wrote in a journal, "Me take charge in a math class? I'm never going to make it! I was very hesitant to even get started (on my own plan), so I put off doing math for a whole week of the course." Because it is difficult for students to accept the confusion that accompanies learning, they need strategies for dealing with uncertainty even more than encouragement to try again.

Although the extent to which school culture discourages self-help is open to speculation, there is little question about the persistence of habitual quick response. Much depends on the latitude students are given, and give themselves, to allocate time for thinking and to examine concepts, relevance, and possibilities. A seventh grade teacher wrote a complex, unfamiliar word problem on the chalkboard and asked students to write the time it would take them to solve it. The average time reported was three minutes. Unfortunately, taking time to solve problems tries the patience of students.[10,11] Too many are prone to answer quickly or abandon problems without attempting to find answers. Rather than rely on their own ability to figure things out, these students want certainty before they proceed and assume that their work is correct without monitoring what they do.

The delivery of instruction itself might be a debilitating factor. One of the disadvantages of an oral presentation is that listeners hear each sentence only once, and expect to learn the lesson as fast as they hear it without further reflection. Failure to learn through this method tends to reinforce feelings of inadequacy. Although a printed text allows for rereading at a selected pace, most teachers do not expect students to learn mathematics through printed explanations. They sense that students prefer to listen rather than take time to reread and analyze.

According to a sixth grade teacher, the quick-answer syndrome is prevalent among high-achieving students who equate achievement with speedy calculations. She worked with youngsters who used algorithms efficiently in familiar, complex situations, but politely rejected thinking that did not produce results in a straight-forward manner. Although identified as academically gifted, these students did not tolerate ambiguity well or credit themselves for making new connections in novel situations. In other groups of highly capable students, teachers create a climate that fosters their natural claim to persistent learning.

Two examples illustrate the common pitfalls of answering immediately without concern for more thought on the matter. In each case, familiar language probably gives students a false sense of security.

Example 1 A potted plant costs ten dollars. The plant costs nine dollars more than the pot. What is the cost of the plant? The pot?

Solution A:	Pot $ 1.00 Plant $ 9.00	Solution C:	Pot $ .50 Plant $ 9.50

Solution B: Pot and plant $ 10.00

Example 2 Write an equation to describe three times as many apples (a) as bananas (b).

Solution A: 3 a = b Solution B: 3 b = a

Unfamiliar questions are likely to induce quick judgments. Asked to illustrate 3 x 18, a student in the sixth grade remarked, "You can't draw multiplication." She did not believe that a drawing could explain why 3 should multiply both digits in 18. After trials and reflection, however, her own illustration of 18 three times was more convincing than symbols describing the distributive property or a drawing she did not construct.

From another point of view, some students assume that working longer and harder should be effective even though the mediating goals and strategies they use are inadequate. As one adult remarked, "I usually tried to be neat in school, work as accurately as possible, and follow the rules; but I never stopped to think or to understand what was going on." Using more time on a problem meant time rewriting papers to make the handwriting neater. Habits like these often stem

3 x 18

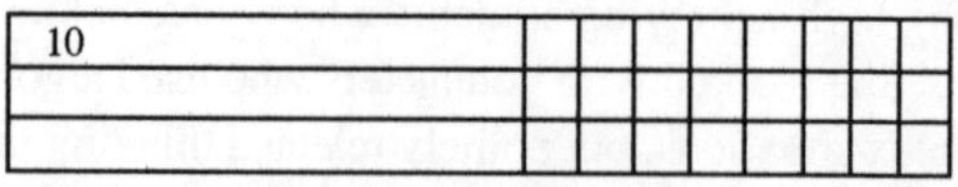

5 x 10 + 4 x 1

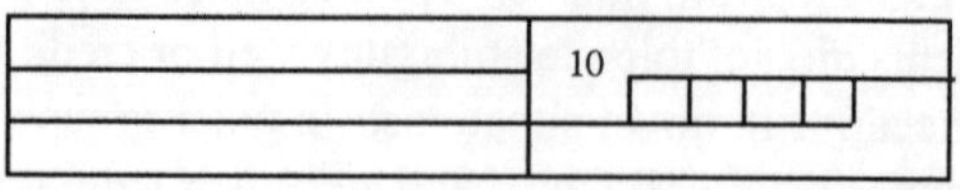

from lack of trust. Young girls and children from families most alienated from school culture, for example, tend to mistrust their own thinking,[12] and alienation is compounded by student assumptions that their own experiences are irrelevant.[13]

For students who have no regard for anything but an answer or to finish an assignment, the relevance of neatness in a drawing is often remote. They might want teachers to read their work, but few expect to refer to their own work weeks later or even minutes into a problem solution. On the other hand, some say that a concern for producing a large number of ideas is more important than neatness. As long as students are aware of the choices they make and test their decisions, they have feedback for directing new choices. Which representations led to new meaning? To new questions? Students can learn to make distinctions among simple recall, their own understanding of concepts, and representations to communicate what they understand.

AUTONOMY

Before children are introduced to academic learning, they use a global approach to learning language and quantitative relationships. A sense of thinking autonomy helps them connect symbols to the world. Students *have a world* to comprehend and reason about if they are able to imagine freely. As students grow older, they tap their experience and produce knowledge to enrich meaning.

Although we do not expect young children to reflect on the difference between theories and evidence for an argument,[14] we do expect them to think in the ways that they are able. Children in the second grade, for example, can create ideas freely and decide consciously to use self-help aids instead of asking for help unnecessarily. They can learn to strengthen their beliefs about the importance of understanding and collaborating, attribute less importance to conforming to the solution methods of others, and less importance to task-extrinsic reasons for success.[15]

Projecting self-direction to its later stages of development, we might expect that children and adults select different learning strategies. Children would choose to manipulate objects and interact with peers as aids to understanding mathematics and adults would use an even larger array of strategies, including the manipulation of objects, drawing, spatial patterns and symbolic equations.

In today's schools, however, such a conjecture is doubtful because students are not invited to generate options or decide how to represent answers. In most assignments and classroom tests, teachers do not offer choices and students do not expect to use self-direction.

Suppose students had a choice between graphics and symbols to describe changes in the slope of a line. Suppose also that students with low self-esteem believed that teachers expected them to choose thinking strategies and representations for ideas. Would students who feel out of control in academic settings direct their own thinking? For some students, social realities prevent them from believing in the learning power that self-direction releases in them. In the typical classroom, the change would be so great that it would take months for the adjustment to occur. Only with significant changes in instruction do individuals control their own thinking enough to make learning a satisfying experience.

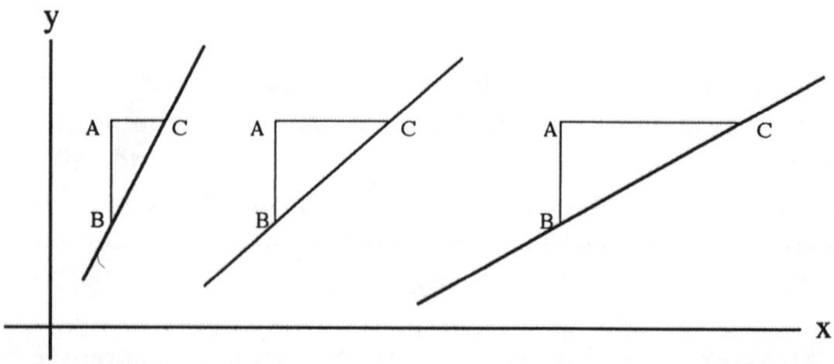

To understand that variations in triangles affect the *slant* of a line, AC, students could select (a) a symbolic expression, $(y_2 - y_1) / (x_2 - x_1)$, or (b) idea production about graphics showing the length of segment AB remaining constant as BC becomes longer. Without self-direction, many revert to formulas because symbols are familiar and suggest certainty. Early and extensive self-directed thinking in mathematics would foster new choices by introducing students to a wider range of thinking strategies and their responsibility for connecting mathematical content. One goal is to have students connect (a) segment lengths, (b) ratios holding the denominator constant, (c) $x_2 - x_1$ as length, and (d) the direction of lines. Another goal is for students to decide when they have drawn enough triangles on the coordinate plane to make connections for themselves.

Older students can establish their own expectations for thinking by examining the ideas listed by youngsters who used free, productive thinking to generate lists of ideas. If children can direct thinking, especially idea production in mathematics, older learners can do it also. Learning through connection-making and self-initiated thinking is open to them in greater measure because they have more experiences from which they can draw.

The selection of thinking strategies is a minor task compared to the challenge of handling quantity and pattern in society without self-regulation. There are adults with home and employment responsibilities who struggle with mathematics anxiety in many of their daily activities.[16] Estranged from the formality of school mathematics, they accept levels of mathematical thinking far below their potential. Adults confident in mathematics have a different challenge. Their self-direction should lead to original problems and unique ways to solve them. In a changing world, students functioning at all levels of achievement must find problems in new contexts, construct meaning for them, and design solutions that they have never seen. They should do much productive thinking without fine-tuned assistance from an instructor.

PROBLEM SOLVING CONNECTIONS

As a practical matter, problems should not be construed as negative experiences or something to avoid. They are opportunities. Problem solving is a general approach for making things happen whether the objective is satisfying school requirements or finding ways to improve life situations. The thinking associated with problem solving contributes to solutions in mathematics like it does in a game of golf and buying a birthday gift in the third grade.

As problem solvers, individuals interpret mathematical content and manage a repertoire of strategies to meet challenges. Performance weighs heavily on accessing learned content and making connections. Generating ideas is characteristic of learning[17] and thinking that students do to understand and solve problems. There is an expectation that students will construct possible models of reality without allowing perceptions of predetermined absolutes restrict their thinking.[18]

Treating mathematics as an ill-structured discipline is a step toward the dispositional and cognitive changes involved in the construction of meaning.[19] According to Robert Davis, the true nature of mathematics involves processes that demand thought and creativity.

> Doing mathematics means confronting vague situations and refining them to a sharper conceptualization; building complex knowledge representation-structures in your own mind; criticizing these structures, revising them and extending them; analyzing problems, employing heuristics, setting subgoals and conducting searches in unlikely corners of your memory.[20]

Only in familiar situations do algorithms and heuristics fit neatly into a structure. For unfamiliar and complex situations, students do not recognize a pattern easily; instead they cope with an inadequate knowledge base for generating inferences and relations not stated explicitly.[21] This is an active role, one that they initiate and monitor. In it, students produce what they know and make new connections to offset error and provide solutions. An evaluation of these ideas helps them respond constructively to problems wherever they occur. Life decisions about careers and the use of drugs depend on idea production and connection-making as much as the construction of mathematics in school.

Self-directed problem solvers not only create a plan to help them organize data and select strategies, they also accept dispositions that support the process. Expecting to experience risk and uncertainty, problem solvers produce ideas because they know their ideas contribute to learning. To avoid false confidence, they look for alternative points of view. The guiding principle is, "The more complex or unfamiliar a situation, the longer I expect to think."

Unfortunately, students stop thinking before they connect what they know with what they can see in given data. Although making this judgment is tenuous for an observer, students can make it for themselves. Incorrect answers are often due to lack of focus and lack of persistence rather than ignorance of mathematics.

Joanne, a serious college student who usually performed well, did not appear to use what she knew about a drawing,. The open-endedness of a given question gave her leeway to use her own ideas; yet the demands for making connections to mathematical concepts remained.

Directions State a question based on the given drawing.
Add facts if you wish.

Joanne's response Add to find the area where the steer can stand.

25 + 50 + 25 + 50 = 150 sq.ft.

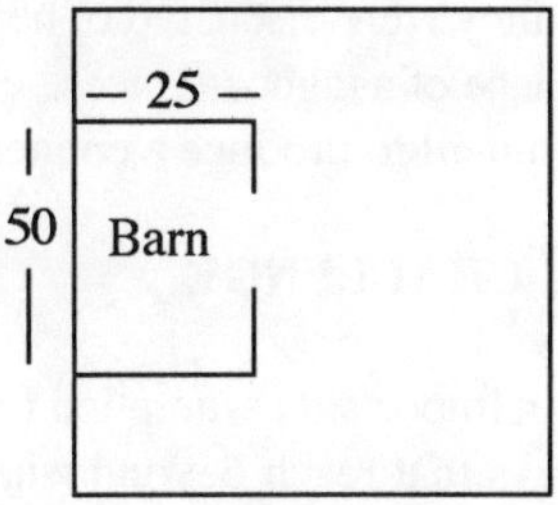

A statement of a question would have provided an objective. Joanne *knew* that 25 + 50 + 25 + 50 feet was the distance around the rectangle (barn). She also knew that steer stand on the region inside the rectangle. Connections between these concepts were missing.

The abstract condition of symbolic expressions takes more connection-making than verbal language tied directly to a familiar context. Students are likely to understand four quarters as a dollar more readily than 4 x .25 = 1. A college student wrote 1 = 1% and drew one box to represent 1%. Although he was familiar with 100% as one complete job, he did not make the connection.

A typical classroom interaction shows that self-directed problem solving is a significant paradigm shift for both teachers and students. In this incident a teacher decided to tell a fifth-grade student what to do rather than develop his informal reasoning and confidence for learning mathematics independently.

Teacher "Where is a right angle?"

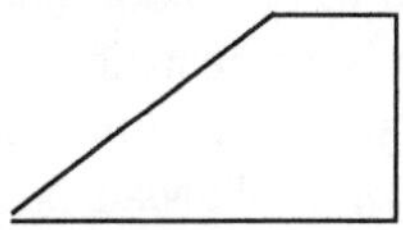

John Points to the center of a piece of tagboard.

Teacher "An angle is the opening between two lines. See the edges of the piece. Straight up-and-down, across. Trace this angle with your finger. Down and across."

Although John traced the angle in the same way as the teacher, he was not using his own resources, other than recall, to do the task. John had not connected the model to prior instruction on the definition of a right angle and its measurement with a protractor. Either he did not internalize right angles on the days they were discussed or he failed to connect what he knew about angles to the shape of a tagboard piece. From the beginning, the teacher did not rely upon the student to produce a connection that was not immediate.

THE CHALLENGE

An important assumption for this work is that students can help themselves in ways that reach beyond what teachers can do. Another assumption is that students and teachers can collaborate to find out what students can do. Because educators do not have fool-proof prescriptions, they depend on input from individual students. The challenge is to change the expectations of students so that they will contribute to their own learning in large measure. With strategies for becoming idea producers, their contributions can be as effective as those of teachers.

My dream is that I could walk into a third grade classroom and ask any child there, "What are you doing now?" The reply: "I'm estimating centimeters to tell me how long the parts of the desk are" or "I'm deciding whether I should make a drawing or just write numbers." This kind of thinking directs the work. It is more useful than, "I want the answer." "I'm almost finished."

A comprehensive program can replace many inhibitors of idea production. Developing self-direction, students become aware that they can study their own experience, connect prior knowledge to new ideas, and propose alternative interpretations. Perhaps most important, students learn that they make the decision to rely upon habit or initiate other strategies. Their own unique thinking is mathematical power that complements direct recall and the voice of authority. Given time and appropriate strategies, students learn that the production of abundant possibilities is more helpful than a treatment of problem solving as a highly restricted activity.

Notes

1. John Dossey et al., *The Mathematics Report Card* (Princeton, New Jersey: Educational Testing Service, 1988). See also U.S. Department of Education, National Center for Education Statistics, *Trends in Academic Progress* (Washington, DC: Educational Testing Service, 1991).

2. National Research Council, *Counting on You* (Washington, DC: National

Academy Press, 1991).

3. National Council of Teachers of Mathematics, *Curriculum and Evaluation Standards for School Mathematics* (Reston, Virginia: NCTM, 1989).

4. James Greeno, "A Perspective on Thinking," *American Psychologist* 44, no. 2 (1989): 134-141.

5. John Goodlad, *A Place Called School* (New York: McGraw-Hill, 1984).

6. Mary Lindquist et al., *Results of the Fourth Mathematics Assessment of the NAEP* (Reston, Virginia: National Council of Teachers of Mathematics, 1990).

7. National Council of Teachers, *Curriculum Standards*

8. Lauren B. Resnick, "Developing Mathematical Knowledge," *American Psychologist* 44, no. 2 (1989): 162-169.

9. Paul Cobb, "Multiple Perspectives," in *Transforming Children's Mathematics Education,* ed. Leslie Steffe and Terry Wood (Hillsdale, New Jersey: Lawrence Erlbaum, 1990), 200-215.

10. Alan H. Schoenfeld, "Teaching Mathematical Thinking and Problem Solving," in *Cognitive Research in Subject Matter*, ed. Laruen B. Resnick and Leopold E. Klopfer (Alexandria, Virginia: Association of Supervision and Curriculum Development, 1989), 83-103.

11. Frank K. Lester Jr. and Joseph Garofalo, *Mathematical Problem Solving: Issues in Research* (Hillsdale, New Jersey: Lawrence Erlbaum, 1982).

12. Lauren B. Resnick, "Developing Mathematical Knowledge."

13. John Brown, Allen Collins and Paul Duguid, "Situated Cognition and the Culture of Learning," *Educational Researcher* 18, no. 1 (1989): 32-42.

14. Deanna Kuhn, "Thinking As Argument," *Harvard Educational Review* 62, no. 2 (1992): 155-178. Other references include Asghar Iran-Nejad, "Active and Dynamic Self-Regulation of Learning," *Review of Educational Research* 60, no. 4 (1990): 573-602; Constance Kamii and G. DeClark, *Young Children*

Re-Invent Arithmetic (New York: Teachers College Press, 1985) on the development of young children; and Mark Johnson, *The Body in the Mind* (Chicago: The University of Chicago Press, 1987) on the embodiment of the imagination in meaning, i.e., metaphorical projections and nonpropositional image schemata.

15. Paul Cobb et al., "Assessment of a Problem-Centered Second-Grade Mathematics Project, "Journal *for Research in Mathematics Education* 22, no. 1 (1991): 3-29.

16. Sheila Tobias, *Overcoming Math Anxiety* (Boston: Houghton Mifflin, 1978). A more intensive investigation than the one given here would treat extensively the effects of social realities like poverty, racial differences, and school administration.

17. Merlin C. Wittrock, "Learning As a Generative Process," in *Learning and Instruction*, ed. Merlin C. Wittrock (Berkeley, California: McCutchon, 1977). Notable examples of groundwork include the research of Reuven Feuerstein, *Instrumental Enrichment; An Intervention Program for Cognitive Modifiability* (Baltimore: University Park Press, 1980) on student generation of new knowledge; Alan H. Schoenfeld, *Mathematical Problem Solving* (Orlando, Florida: Academic Press, 1985) on student beliefs; and Arthur Whimbey and Jack Lochhead, *Problem Solving and Comprehension* (Hillsdale, New Jersey: Lawrence Erlbaum, 1982) on collaborative problem solving that helps students use what they know.

From a different perspective, other studies reexamine the roles of analogical reasoning in learning for children and adults. See S. Vosnaidou and A. Ortony, eds., *Similarity and Analogical Reasoning* (Cambridge, England: Cambridge University Press, 1991). This work includes self-direction insofar as it treats individual plans and goals, and thinking that students initiate, e.g., the identification of surface features as cues to underlying structures.

18. Ernst von Glasersfeld, "An Introduction to Radical Constructivism" in *The Invented Reality*, ed. Paul Watzlawick (New York: W.W. Norton, 1984).

19. Lauren B. Resnick, "Treating Mathematics As an Ill-Structured Discipline," in *The Teaching and Assessing of Mathematical Problem Solving*, vol. 3, ed. Randall I. Charles and Edward A. Silver (Lawrence Erlbaum and the National

Council of Teachers of Mathematics, 1989).

20. Robert B. Davis, *Learning Mathematics: The Cognitive Science Approach to Mathematics Education* (Norwood, New Jersey: Ablex, 1984).

21. M.T.H. Chi, Robert Glaser and E. Rees, "Expertise in Problem Solving," in *Advances in the Psychology of Human Intelligence*, vol. 1, ed. Robert J. Sternberg (Hillsdale, New Jersey: Lawrence Erlbaum, 1982), 7-75.

2 *A Program for Self-Direction*

Routine thinking is often considered more efficient than self-direction because it is predictable, thereby giving teachers greater control over student response. To promote problem solving and meaningful learning, on the other hand, a classroom management system must allow students increasingly to operate as self-regulated and risk-taking learners rather than obedient charges.[1] In this environment teachers transform their expectations from predictable responses to a wide range of thinking strategies that students initiate.

A proposal to develop fluent idea production contrasts greatly with the routine thinking that dominates most schools.

Routine Thinking	**Fluent Idea Production**
Literal Reproduction	*Clarity*
• Reproduce facts, questions. Look for familiar ideas.	• Reread. Seek understanding.
• Use only memorized steps.	• State ideas in my words. Produce alternative interpretations.
One Chance	*Formats*
• Expect immediate results.	• Take time: speak, draw, write.
• Read data quickly. Aim to complete the task.	• Record for future reference.
Fragmented Ideas	*Connections*
• Recall a fact to accomplish the task.	• Listen to words. Use long-term memory. Produce connections original to you.
• Stop thinking when there is an answer.	• Relate conditions and results in different ways.
• In school, use only routine thinking.	• Connect thinking inside and outside of school.

Problem solvers select routine thinking for familiar simple tasks and a variety of productive strategies for challenge problems and the construction of meaning. If people know what to do, they produce an answer immediately. If they do not know what to do, they produce many ideas in search of new connections. A case is made for fluent thinking as opposed to the production of one idea. Ordinarily, common ideas usually come to mind first. Then problem solvers tap new categories and original thinking. This development is fertile ground for studying possibilities, pulling ideas together, and for finding a solution.

When problem solvers sense that routine thinking is inadequate, they turn to self-directed, fluent idea production. Susan, a high school graduate, recognized that the purchase of her first car was an unfamiliar task. Her school experience had organized data for stylized word problems and taught her to compute. For a situation as complex as a car purchase, however, Susan had not been coached on specific procedures for identifying and comparing cost, payment plans, mileage and other factors that change each year. She can get someone else to decide which car to purchase or gather information and make the decision herself. If Susan takes the latter course of action, she begins with a belief that *I don't know what to do*. Then she produces possibilities and connections in search of a plan. The process is messy and progress in the early stages uncertain.

The same kind of self-direction occurs when students confront unfamiliar computation, concepts, words problems, representations, and other mathematics in school. Where self-direction flourishes, students are aware of what they do to construct meaning and use opportunities. It relies on the *self* that influences both children[2] and adults when they ask questions that do not have the usual form of a *correct answer*. "How can people live in the street?" "Why does my father do that?" It is a moment when they are in charge. As one child put it, "That's *me* thinking. I am myself." Mathematics is not exempt. Confronted with ambiguity, individuals depend on "me thinking" more than external direction. Educational settings are comprehensive enough for both resources.

Programs that emphasize *me thinking* cultivate internal resources. In them, students learn to say what they know, find new connections and choose when to think broadly and when to focus on a conclusion. Able to produce ideas from many points of view, students become problem solvers who construct knowledge and find solutions to new-found problems. Idea production fills gaps and makes knowledge both useful and satisfying. Students also generate ideas when they read. Reading mathematics, it is necessary to supplement the text with recall and invention. This productive thinking makes reading a resource that students can use independently. The dearth of text paragraphs in mathematics textbooks indicates that reading has not been an important resource for students. With self-directed idea production, this situation could change.

PROGRAM COMPONENTS

SELF-DIRECTED MATHEMATICS (SDM) is a program designed to engage students in active problem solving from elementary school through college. In this program, productive thinking is the catalyst that begins problem solving and sustains it in novel and complex situations. Self-directed idea production permeates the construction of mathematical concepts as well as plans for responding to a dilemma outside of school.

Six components integrate the program.

- ***MATHEMATICS*** Students construct meaning and problem solutions for a wide range of challenges. Current local and national recommendations direct the choice of mathematical topics and instructional strategies.

- ***SELF-DIRECTION*** Students develop a sense of autonomy that helps them initiate productive thinking, examine their beliefs, and use their differences to advantage. They read mathematics, allocate time for thinking, produce alternative interpretations, and make multiple connection among ideas, drawings and contexts. (Chapter 2, 3).

- ***IDEA LISTING*** Students produce ideas freely to tackle novelty and complexity, to clarify concepts, and create problem solutions. Resources include recall, observation, imagination and peer interaction. A checklist guides their thinking into mathematical concerns and informal prompts develop a climate for problem solving (Chapter 4).

- ***PROCEDURE*** For challenge problems, students (a) list many ideas about given facts; (b) restate questions to insure meaning; (c) list many ideas in drawings, words and symbols that "might" lead to solutions; and (d) select their best ideas. They solve given word problems and those in which they add facts and a question. A modified version of this plan helps them understand computation and respond to social situations (Chapter 5, 6).

- ***MATERIALS*** Diverse materials and technology stimulate thinking and multiple representations (Chapter 3).

- ***ASSESSMENT*** Measurement of productive thinking described in Components 3 and 4 reaches beyond most conventional instruments and supports self-directed problem solving (Chapter 7). Portfolios and other on-going assessments show the development of thinking strategies, dispositions, and mathematical knowledge. Assessment includes progress on self-help.

Under self-direction, students initiate action as well as follow directions, and state individual objectives as well as follow the recommendations of teachers. As a general rule teachers do not do anything that students can do for themselves. Instead, they ask students to look for alternatives and design new ways to use their abilities and opportunities along with many traditional approaches to learning and problem solving. Participating as thinkers and problem solvers, students produce ideas as they listen, read, write and draw to make sense of new concepts and applications.

Confronted with a challenge, students switch from *what I have done in routine situations* to *what I construct*. This is a choice of self-direction over automatic response, and of independent thinking over an inordinate dependence on *what I have been told.* With strategies and dispositions for solving problems, they are prepared to act in social situations as well as academic learning.

Teachers should be aware that ***SDM*** components do not serve the purpose intended for them well when elements of other programs inhibit idea production. It is helpful for teachers and students to state their expectations: "This activity requires routine thinking with a given time limit." "The assignment requires fluent and original idea production." Most students are flexible enough to adjust their own thinking and select strategies for situations themselves.

STRATEGIES

SDM emphasizes the importance of independent thinking and the ways in which students help themselves. Strategies help students integrate new experiences in mathematics with their own prior knowledge, test the integration and make adjustments to increase understanding and invention. These are problem-solving functions that rely upon self-direction rather than routine. When students understand this, *self-directed problem solving* becomes a redundant expression.

Choosing to think and claiming the authority to produce ideas are commitments that rely on the development of metacognition, the awareness of mental functions and executive decisions about when to use them.[3,4] Metacognitive functions help students regulate (a) cognitive operations, e.g., recall, infer, and compare; (b) strategies, e.g., draw and list possibilities; and (c) metacognitive action to plan strategy, monitor it, and allocate time for thinking. These functions emphasize the production of ideas unique to individuals as a supplement to the directions of teachers. Reaching into cultural differences[5] and individual knowledge gaps, metacognition contributes to learning transfer, academic curiosity, and persistence.[6]

The program refers to *self-direction* instead of *metacognition* to emphasize that students are the ones to examine and initiate cognitive strategies. Experience with self-direction helps them recognize their part in the development of mathematics especially in the enrichment of their own understanding. Monitoring thinking, students decide when to probe long-term memory and when to combine ideas into new inventions. An active assimilation prepares them to elaborate on what they know and develop representations that communicate their knowledge and problem solutions to others. Implications extend to what students believe about mathematics and themselves.

SDM activities center on an autonomous production of ideas in a search for new connections. The word, *idea*, suggests that students expect to produce possibilities for answers or for the direction of complex problem solutions. A new connection is an idea. Whereas *answers* are to be correct, the immersion of *ideas* into subject matter[7] introduces problem solving that encourages growth and revision. "What is your idea, Susan?" "List many possible ideas." Responses to these questions are a student's contribution to an exchange of ideas, and the discourse created builds understanding.[8] Where expectations for a uniform development of meaning do not interfere, students take intellectual risks that reach beyond minimal prescriptions and perceive that an extensive generation of ideas is as much a part of school performance as the reproduction of definitions and algorithms.

Producing prior knowledge and new inventions, students unify their consciousness of facts, questions, and solutions. When understanding is not forthcoming, they use a flexible production of ideas that depends initially on total acceptance from teachers and peers. Each idea fosters the flow of ideas and contributes to better ideas. Self-direction and understanding reinforce positive attitudes toward thinking so that students face novelty and complexity with a sense that *I can think* and *I always know how to begin.*

DIFFERENCES

The core of the ***SDM*** program is thinking strategies that become resources directed by individuals. Among the many factors that support productive thinking and individual control are student regulation of the pace of learning,[9] activities that require on-going student response, and interaction with peers. More than a delivery system for learning subject matter, the program develops the construction of meaning and problem-solving strategies to be used throughout life as well as in school mathematics.

To develop insight in school and to transfer knowledge to new contexts, students need to manage thinking consciously and uniquely. In this sense they share the direction of thinking with teachers.

- For me to construct ideas, I must be in charge of my own thinking.

- For me to use *my* uniqueness to do *your* mathematics, I need to monitor the learning strategies I use.

We know little about what students are able to do in this domain, especially through a deliberate cultivation of self-directed thinking. We do know, however, that students are in the best position to direct and monitor their own thinking. They can choose to recall and connect ideas from diverse sources, past and present, and add *what I figure out* to *what you expect me to recall.* As individual authority for thinking develops, self-direction and ownership in the learning process lead to unique accomplishments as well as those prescribed in a course of study. The process helps individuals gain the maximum use of their own mathematical power by supplementing classroom activity with self-directed thinking.

Regulating their own habits, ***SDM*** students become aware of factors that influence motivation, namely, degrees of challenge, curiosity and control.[10] Self-direction addresses challenge by prompting individuals to stretch and monitor their own competence. Too little challenge diminishes motivation and too much challenge makes a task impossible.

In a computer game the uncertainty of the challenge breeds curiosity. The more uncertainty the more curiosity, especially when the responsibility for reducing uncertainty rests with the person engaged in the game.[11] The correlation changes, however, for academic encounters considered outside a student's responsibility for reducing uncertainty. Here students often correlate uncertainty with difficulty. If students favor immediate success over challenge, curiosity and imagination diminish accordingly.

WORD PROBLEM A $200 suit is discounted by 20%. What is the cost?

Student A Some day I want to buy a suit. Maybe I should learn this.

Student B I will figure this out later, when I need to buy a suit.

Student C I like jeans, not suits. The problem does not interest me.

Control extends to the thinking power students are willing to exercise and that teachers allow. Students who make decisions freely at home are constrained by the restrictions incumbent upon them in school. Both youngsters with little home supervision and adults who manage a household see themselves as directors of their own activities. For some, a simple requirement that everyone in a class must do the same thing at the same time is an unreasonable expectation and their aberrant behavior lets teachers know that they want to be in charge.

In other groups of students, cultural experiences have nurtured helplessness. Repression at home and overstated respect for educational experts are but two of the factors that dispel self-help. Parameters for self-direction give these students a base for self-confidence and a license to contribute to their own learning.

A more intensive investigation than the one given here would treat the effects of social realities like the poverty and race much more extensively. Students have different responses to mathematics because they differ in prior knowledge, experience, and disposition. To make the matter more complex, large numbers of students work with teachers who have an economic status or cultural heritage unlike their own. Teachers who have had little experience with the working habits and values of another culture are particularly dependent upon leads from students.

Attitudes differ in persons of the same race, including Caucasians, and expectations for teacher direction vary considerably. Some African-American cultures mistrust teachers while Asian cultures are more likely to reinforce teacher dominance. In a program that encouraged minority students to attend college, for example, low-achieving sixth grade students decided that mathematics was not the thing to do on Saturdays or during the summer. After a few class sessions, one student set the tone by declaring that the group would not use paper and pencil. For this group, Cuisenaire rods and computers not only provided a change of pace but also called upon different student abilities and greatly improved interaction between students and the teacher.

How does a teacher balance the impact of culture on students with recommended strategies for helping student learn? How do teachers know which cultural influences prevail? Although in the most situations teachers are expected to adapt instruction to groups, they do not readily create the best prescription for each student. The ***SDM Program*** responds to this challenge by inviting individuals to manage their own thinking far beyond what teachers can do to help them. In addition to following the prescriptions of teachers, students find other ways to understand and do mathematics that makes sense to them.

The conventional language of mathematics instruction is a concern because it differs from everyday discourse[12] especially for students growing up with backgrounds that do not favor technical exposition. Writing word problems themselves, students use their own language and describe their own interests. For example, children in elementary school designed these problems:

> Antonio had 7 jars with 10 ants in each jar.
> He put 17 ants in his parents' bed and 30 ants
> in his brother's bed. How many did he have
> left to bake in cookies?
>
> A restaurant lets one child eat free with two
> adults who pay for meals. A club has 30 adults.
> How many children can eat free?[13]

Differences extend also to representations of mathematics. Students who like to write, for example, often prefer to express their ideas in essays and other forms of composition. Many claim that writing, speaking aloud or discourse helps them think. They need to determine whether (a) written verbal expression is, for them, a prerequisite for understanding graphic and symbolic representations or (b) drawings and equations are a greater help for synthesizing ideas from paragraphs. Coming to terms with this question can deter negative perceptions of self and mathematics.

Why students depend commonly on verbal expression is debatable. Are students verbal because they have had little experience with drawings or does ability influence a decision to depend on words? Do some students with spatial strengths hide them in deference to convention? Five groups of students in the fourth grade, each at different rural and city schools, were asked to list as many ideas as they could about fractions. Most students wrote words, and only words. For them, writing *about fractions* did not include examples in numbers, especially in fractions greater than "one" or with large numbers. Drawings were even less likely. If there was a drawing, it was a pie-shape that students had seen in a textbook. To encourage divergent thinking, the teacher asked two children to write *a funny fraction.* With some laughter at the suggestion, they listed 30/1, 4/3, and 17999/17.

The preponderance of verbal expression and brevity applies also to lists about fractions generated by college students. Although students taking advanced-level mathematics courses include symbols and drawings more freely, they commonly confine their lists to recall items and original thinking is rare. Many students without a strong mathematical preparation tend to think about concepts verbally before they include representations other than words. Without practice, few students trust themselves enough to show their actual experience with mathematics.

Which drawing shows 1 / 3 x 1 / 2 = 1 / 6 ?

A B

In Drawing A the entire left-hand column shows 1/3 of the rectangle. Drawing B shows only 1/3 of 1/2 of the rectangle. Claiming that one drawing is as good as the other fails to distinguish between a mechanical manipulation of symbols and an interpretation of 1/3 x 1/2 as part of 1/2.

Perhaps few students see themselves as problem solvers concerned about the meaning of multiplication of fractions. They can learn to construct meaning when they see "x" between fractions. Some might associate it with *times* as *⅓ of one time they see 1/2*. They will learn other meanings for the symbol as they learn more mathematics. To focus their work, students ask, "What am I trying to do?" "Does my drawing show what I intend?" "Do I believe what I wrote?" "Does this drawing clarify the situation or mislead?"

BELIEFS

Dealing with complex problems and definitions with procedures used by experts in a routine fashion discourages thinking about the nature of mathematics and specific situations. Without self-direction and a search for alternatives, there is a tendency to abandon problems or use faulty heuristics with conviction. Six college students, for example, insisted that *1/2* refers to two parts, but not necessarily parts of the same size. One part could have a larger area or include more items than the other. Strangely enough, students actually believe something other than what they say, and then act on this belief.[14] They might quote the appropriate rule about dividing by zero, but actually believe that 3/0 = 0. Some use procedures that they do not believe. "I know it's wrong, but I thought you wanted it that way."

In a typical response from twenty-five students in the seventh grade, only two students found the correct solution to a problem. Students generally knew the required mathematics and had ample time to work on the problem.

> **Problem** John Troy owns a full line of farm machinery. He decided to change oil in his tractor, combine, large tractor, grain truck, and pickup. The large tractor uses 12 quarts during an oil change, the combine uses 4 gallons, and the garden tractor uses 1/4 as much as the large tractor. Both trucks use the same amount of oil during an oil change, 3 quarts less than the large tractor.
>
> If John bought all the oil at the same time, and it costs $.94 per quart, how much change would John receive from a $100 bill?

Most students listed ideas sparingly. This choice made heavy demands for the memory of variables and conditions. Errors included *4 quarts + 4 gallons = 8 quarts* and reference to only one truck.

We can only guess the influence of beliefs that students have about mathematics and about their own thinking power. What do they believe about the effects of reading carefully and the time it takes to do mathematics? How do they help themselves understand what they read and what do they do to connect answers, given facts, and the question asked?

An activity, not usually associated with mathematics, dramatizes the tendency of individuals to underrate both their experience and their ability to produce ideas. Participants were groups of teachers, business people, and elementary school students who were asked to write an estimate for the number of birds (or fruits and vegetables) that they could name in five minutes. Then they wrote the names of birds. In more than fifty groups, about two-thirds of each group found that they could name far more birds than they anticipated. Asked to consider categories of birds, e.g., color and size, they made their lists even longer. Attention to the simplicity of the directions given, *to name birds*, led to responses like *Larry Bird* and *jailbird.* Participants concluded that they had restricted their responses unnecessarily.

Claiming that mathematics is predetermined, students produce ideas with less fluency and flexibility in mathematics than on other topics. They list ideas about rectangles, for example, without naming the four sides and internal/external regions that they see. They do not make a figure with a length six times longer than the width or with sides not parallel to the sides of the paper because they do not believe that rectangles could look that way.

The ***SDM Program*** seeks to engage students in an examination of their beliefs about idea production, authority, and mathematics. Individuals challenge their beliefs to minimize the interference of habit with the construction and application of knowledge, and anticipate that beliefs change as they synthesize more ideas and learn to develop new points of view. At the same time, students act upon what they believe unless they see a contradiction. Rather than assume they have little to contribute, students transfer the rational beliefs and strategies used outside of school into the classroom.[15,16]

The program not only urges students to choose self-help but also provides a wide range of strategies to help them do it. Students understand that they choose to read or skim passages and that they select drawings,words, and symbols. Their choices direct learning and problem solving.

To be effective, ***SDM*** must establish new roles for students and teachers. Roles for both groups include decisions in nearly all aspects of curriculum and the activities designed to promote learning. It will take many years before we know whether self-direction and idea production help students construct ideas and influence the transfer of learning between school and out-of-school contexts. Then students will be aware of their thinking and be free enough to describe the choices they make to enhance problem solving.

Notes

1. M. McCaslin and T.L. Good, "Compliant Cognition: The Misalliance of Management and Instructional Goals in Current School Reform," *Educational Researcher* 21, no. 3 (1992): 4-17.

2. Robert Cole, *The Spiritual Life of Children* (Boston: Macmillan, 1990).

3. John H. Flavell, "Metacognition and Cognitive Monitoring," *American Psychologist* 34, no. 10 (1979): 906-911.

4. Robert J. Sternberg, *Intelligence Applied* (San Diego: Harcourt Brace Jovanovich, 1986).

5. Diana T. Slaughter, "Black Children, Schooling, and Educational Interventions," in *Black Children and Poverty: A Developmental Perspective,* ed. Diana T. Slaughter (San Francisco: Jossey-Bass, 1988).

6. Scott G. Paris and Peter Winograd, "How Metacognition Can Promote Academic Learning and Instruction," in *Dimensions of Thinking and Cognitive Instruction,* ed. Beau F. Jones and Lorna Idol (Hillsdale, New Jersey: Lawrence Erlbaum, 1990).

7. Richard Prawat, "The Value of Ideas: The Immersion Approach to the Development of Thinking," *Educational Researcher 20*, no. 2 (1991): 3-10.

8. Magdalene Lampert, "Choosing and Using Mathematical Tools in Classroom Discourse," in *Advances in Research on Teaching*, ed. Jere Brophy (Greenwich, Connecticut: JAI Press, 1989).

9. M. Ann Dirkes, *Producing Ideas in Mathematics Manual* (Buffalo: United/ D.O.K. Publishers, in press).

10. Jill H. Larkin and Ruth W. Chabay, "Research on Teaching Scientific Thinking: Implications for Computer-Based Instruction," in *Toward the Thinking Curriculum: Current Cognitive Research,* ed. Lauren B. Resnick and Leopold E. Klopfer (Alexandria, Virginia: Association of Supervision and Curriculum Development, 1989).

11. A. Wade Boykin and Judith Harackiewicz, "Epistemic Curiosity and Incidental Recognition in Relation to Degree of Uncertainty: Some General Trends and Intersubject Differences," *British Journal of Psychology* 72 (1981): 65-72.

12. Richard Cocking and Jose P. Mestre, eds., *Linguistic and Cultural Influences in Learning Mathematics* (Hillsdale, New Jersey: Lawrence

Erlbaum, 1988).

13. Arthur Whimbey, Jack Lochhead and P. Potter, *Thinking Through Math Word Problems* (Hillsdale, New Jersey: Lawrence Erlbaum, 1990).

14. Alan H. Schoenfeld, "Beyond the Purely Cognitive: Belief Systems, Social Cognitions, and Metacognitions as Driving Forces in Intellectual Performance," *Cognitive Science* 7 (1983): 329-363.

15. Alan H. Schoenfeld, "On Mathematics as Sense-Making: An Informal Attach on the Unfortunate Divorce of Formal and Informal Mathematics," in *Informal Reasoning and Education,* ed. J. Voss, David N. Perkins and Judith Segal (Hillsdale, New Jersey: Lawrence Erlbaum, 1991), 311-344.

16. Paul Cobb, "Multiple Perspectives," in *Transforming Children's Mathematics Education,* ed. Leslie Steffe and Terry Wood (Hillsdale, New Jersey: Lawrence Erlbaum, 1990), 200-215.

3 *The Power of Self-Directed Thinking*

According to Frank Smith,[1] we are forever trying to give meaning to words that name aspects of our experience. Thinking is a sea of interaction among assumptions, inferences, and the search for clarity.[2] Sensing ambiguity, novelty or complexity, we use these interactions to make decisions about what we believe and what we will do. Rather than abandon a task or ask someone else for direction, students need to convince themselves that they can use their own idea production to meet challenges. Thinking is their greatest resource, and self-directed thinking is mathematical power.

The ***Self-Directed Mathematics Program (SDM)*** emphasizes constructive thinking in a fluid model for solving problems. It sustains the natural ability of students to act as problem solvers and develops new strategies that make self-direction an integral part of learning and doing mathematics. The program makes students aware of their own growth in mathematical power and increases self-help to maximize benefits from school activities. In this sense, students form a coalition with teachers to construct meaning and make decisions about ways to help themselves.

To develop thinking initiative, ***SDM*** makes self-direction an instructional priority. As in every growth exerience, initiative responds to the challenge of new situations and various gaps in learning. Because individuals bring their own history and ways of doing things to a task, they are the ones who direct their unique abilities and resources to the best advantage. Three levels of strategies contribute to mathematical power.

Strategies for Self-Direction

Entry Level Teacher initiated, student executed

Follow brief written and oral directions.
Make choices among **given options**.
Ask for nonspecified "help."

Respond to **prompts** for using idea production and other strategies.
Use a **guide** to self-help.
Accept **uncertainty**.

Use reference **books** and other physical resources.

Intermediate Level Student initiated

Specify the help required. "**Please repeat** what you said."

Reread for the main idea and for details.
Interpret directions and concepts that pose a dilemma.
Restate questions and objectives in familiar vocabulary.

Speak alone or in discourse with teachers and peers.
Ask pertinent questions to increase understanding.

Record ideas produced outside of school.

Select the length of **time** for thinking and the appropriate degree of accuracy.

Ask, "Am I **challenging** myself?" "What do I believe?"

Advanced Level Student initiated and combined with intermediate strategies

Read and follow directions written for conceptual learning without an oral introduction/explanation.

Elaborate on given data and answer in variable responses.
Make drawings and charts that solve problems.

List original and alternative ideas on paper. Compare.
List multiple connections; categorize and synthesize ideas.

Create a context when none is given.

Select, create and monitor **thinking strategies.**

Find problems: opportunities, dilemmas, challenges.

Use a problem solving **plan** with extensive idea production: state facts, the problem or objective, possible solutions, and check on agreement among facts, questions and answers.

Although students learn some strategies during the usual classroom experience (e.g., reading for detail), ***SDM*** adds requirements to prolong the time used for thinking and to monitor the effects of the strategies used. Evaluation of self-direction includes (a) the novelty of the task, (b) prior mathematical knowledge, (c) personality variables, and (d) the time and freedom students have had to work on self-direction in school.

ENTRY LEVELS OF SELF-DIRECTION

In the early stages of development students become aware of self-directed thinking, especially the strategies they invent themselves. This is a period in which individuals begin to use a license for producing ideas. Teachers ask students to make choices and to help themselves. "Say what you know." "Ask me a question." "Guess two things and choose one." There is a clear distinction between "What is the answer?" and "What can you say about the situation?"

Other prompts stretch thinking further. "What else do you know?" "What else might be important?" "What do you want to revise." "What can you do to help yourself?" "You can decide." Although the last statement often refers to management decisions appropriate for students, the habit of making decisions carries over to competence in mathematical thinking.

Question What makes 45? Produce many ideas.

Rather than assume students will identify a particular response, like $5 \times 9 = 45$, students search for many possible connections. The question prepares them to respond to ambiguity and find solutions to problems.

As a general rule, students find out what they can do rather than depend on a teacher to tell them. A child who decides to make a drawing that shows the renaming of ones as a ten, for example, says more about our number system and uses more self-direction than an adult who simply writes the sum of 3, 5 and 8 through the recall of addition facts. Self-directed searches for an understanding of the number system are likely to predict future performance in mathematics.

A guide to self-help bridges the gap between prior experiences with learning as a highly dependent process and learning influenced by student idea production. It reminds students to tap their own resources before asking for help.

Guide to Self-Help

1. Read it again.
2. Draw and write.
3. Make many guesses.
4. Talk about it.
5. Produce more ideas.
6. Say what you are doing.
7. If 1-7 is not effective, ask a teacher.

After an orientation to the guide, teachers post it to remind students to seek understanding through careful reading and idea production. Drawings and many guesses (ideas) help them produce answers. When students say what they are doing, they focus on the task at hand.

Most classroom activities in the ***SDM*** program involve students in projects that develop concepts and problem solving. There is an assumption that while a preoccupation with the literal interpretation of directions interferes with knowledge construction and independent thinking, productive thinking nurtures problem solving. Student expectations for flexible idea production foster a conviction that *I can find meaning and act upon it*. The need for external affirmation decreases as students begin to evaluate their own progress in self-direction as well as their understanding of mathematics.

A preservice teacher related an incident in which she helped a student comprehend what he read. Rather than repeat directions, she encouraged the student to counter ambiguity with a trust in his own thinking.

> Today while I was teaching a lesson in fifth grade, a student asked me how to do an assignment. We had just gone over the directions and I had given an example. So I simply asked, "What do you think you should do?" He said, "I don't know." (Those famous three words.)
>
> I asked him, "What *do* you know about this page?" From that, he went on to talk about the mathematics and we got to the point where he was telling me what he was to do. I said, "See, you did know what to do." He smiled.

When teachers use an unfamiliar format, students tend to ask, "What do you mean?" The question usually indicates a preference for certitude over intellectual risk. To construct mathematics, however, students need to explore and imagine. The beginning of a response or a few guesses contribute more to the construction of sense-making for the student than more information from a teacher. The more students talk the more they clarify ideas for themselves.

INTERMEDIATE LEVELS OF SELF-DIRECTION

Students move to intermediate levels of self-direction by using strategies without a cue from someone else. At entry levels teachers ask students to read for details and restate questions in their own words. At intermediate levels students are the ones to initiate these strategies. In addition to self-directed comprehension strategies and the selection of supplementary materials, students regulate time, accuracy and interactions with peers and teachers. The initiative and the strategies developed prepare them for advanced levels of self-direction.

Prolonged idea production and self-monitoring are precautions against faulty schemata. Unless students produce possibilities at their own direction, they cling to error and limit their understanding of concepts. Some students assume, for example, that buying three items for a dollar means that they pay a dollar for each of the three items. Asked to produce more ideas about the concept, they might infer that an advertisement, *3 for $1.00*, should read *1 for $1.00*. Stretching further, they question their original beliefs about *3 for $1.00*.

Another sign of heightened self-direction is the conscious interpretation of a question. Two students in the third grade interpreted directions for a mathematical activity that was a novel experience for them. The problem required the largest sum found by going from Point A to Point B.

Tom added numbers on a direct route from A to B: 3 + 4 + 4 + 2. Hank claimed that the directions did not say that the route must go *straight from A to B*. He found a larger sum by adding all of the numbers in the drawing. Explaining that answers depend on an interpretation of the question, the teacher accepted the ideas of both students. At this point she heard Tom say to Hank, "Hey, that was neat the way you did the problem. I never thought of it."

Figure 1 The longest sum

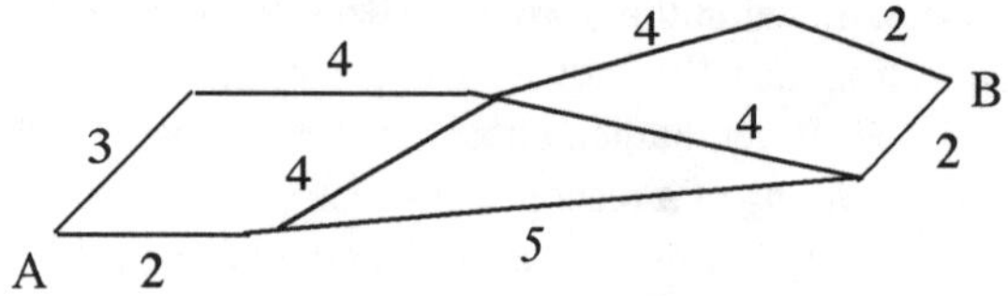

In this example, Tom recognized the intent of the mathematician who wrote the problem. Hank sensed a creative interpretation of moving from A to B. At the same time, he might have enjoyed surprising others with an unusual response or recalled a decision he once made to take the *long way home* in order to delay an unpleasant experience.

Restating the objective clarifies the intent.

Textbook Question How far did Mr. S. travel?

Restatements
How far did he go?
How many miles was it?
How many hours estimated for the trip?
Find the number of miles traveled.

In each restatement, students not only search for meaning, but also select units (e.g., miles) and establish a direction for their work. The number of restatements usually increases for complex situations and for questions unfamiliar to the problem solver.

Computation and algebraic manipulation call for similar interpretations of the question. To solve for y in $xy + 3x - y = 2$, many students begin by adding y and -2 to the equivalent expressions.[3] In itself, this step is acceptable but does not move toward the objective. Restating the question, however, has a positive effect on comprehension. What does it mean *to solve for y*? What does it mean to me? *Find all of y* or *Get all of y as soon as possible*.

In a familiar context, there is a natural interplay between knowledge construction and self-direction. Experience with grocery shopping and building cupboards, for example, brings purpose and detail that are absent from academically styled problem solving. Similarly, adults in the work place tend to estimate warehouse space through comparions of the number of items that fill it rather than calculate cubic feet.[4] Recording the ideas they produce outside of school, students increase their confidence in productive thinking.

ADVANCED LEVELS OF SELF-DIRECTION

Student-initiated idea production dominates the advanced levels of self-direction. When questions are obscure or recall inadequate, students produce ideas that they have not seen or do not recall. Ideas are connections produced as students read mathematics, draw and do original thinking. At advanced levels, young students as well a older ones find challenging problems and select strategies to solve them. Fluent responses in a variety of representations facilitate their work.

READING

To learn mathematics by reading print materials or a computer screen, students produce ideas abundantly in a search for comprehension. They make their knowledge interact with a text in the construction of meaning[5] without a teacher's oral explanation. Interactions occur as they (a) state an objective; (b) connect words and symbols to a drawing; (c) produce inferences and other connections; and (d) compare work with the objective.

Although initial responses are often awkward and even incorrect, the combination of reading, drawing, and idea production prompts reflection that leads to a belief in mathematics and in personal efforts to use it well. Confidence increases as teachers reassure students that their ideas show thinking, and persistence leads to better ideas. The intent is to make individual thinking replace the inactivity that often accompanies confusion. When students trust their own reading and thinking, understanding becomes a driving force.

Reading activities[6] can be designed to develop understanding through productive thinking. Spatial representations embody algorithms and activities end with a thoughtful review of a concept or put computation in a context. Students use fluent idea production to answer many questions and teachers look for student ideas as well as correct answers.

Reading Activity 12 cm = 1 dm 2 cm = 1.2 dm

1 a) Draw 12 boxes to show 12 ones. 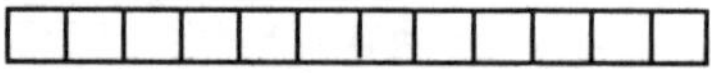

b) Show 1 ten and 2 ones. 

10 2

c) Write: 12 boxes = 10 + 2

2 a) Draw 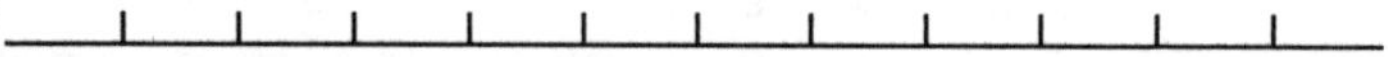

— — — — — — — — — — — 12 cm — — — — — — — — — — — — —

b) Draw 

— — — — — — — — — 1 dm — — — — — — — — — — 1 cm 1cm

c) Trace a line segment 10 cm long.
Trace a line segment 1 dm long.

d) Write: 10 cm = 1dm
10 cm = 10 / 10 dm = 10 out of 10 parts
We can trade 10 cm for 1 dm and 1dm for 10 cm.

3 a) Draw 1 dm 3 cm. Is it the same as 13 cm?

b) Write: 1 dm 3 cm = _____

c) 1.3 dm is one name for the length. Look at the drawing.
Write 3 other names. Use cm, +, and other ideas.

4 a) Draw: 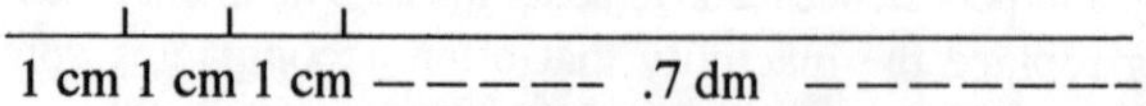

1 cm 1 cm 1 cm — — — — — .7 dm — — — — — — —

b) Point to 3 out of 10 parts. Write: 10 cm = 1dm
3 cm = 3 / 10 dm = .3 dm

5. Measure these segments. Write the lengths.

a) ____

b) ______________________________________

6. Make a drawing to show 2.1 dm. Trade 10 cm for 1 dm.

Tell why the drawing shows:

a) 21 things   b) 2 dm 1 cm   c) 2 1 / 10   d) 21 cm

7 a) How are centimeters and decimeters like ones and tens?

b) How does a centimeter differ from a decimeter?
List many ways.

Reading Activity Averages

1. To find *about how many*: Number of Boxes

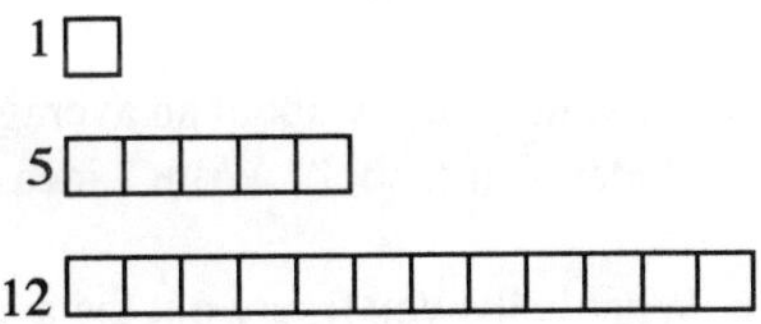

a) Show 1 box, 5 boxes, 12 boxes.

b) Add: $1 + 5 + 12 = 18$

c) See 3 groups. Divide by 3.
$18 \div 3 = 6$
Think of 3 *equal* groups, 6 in each.

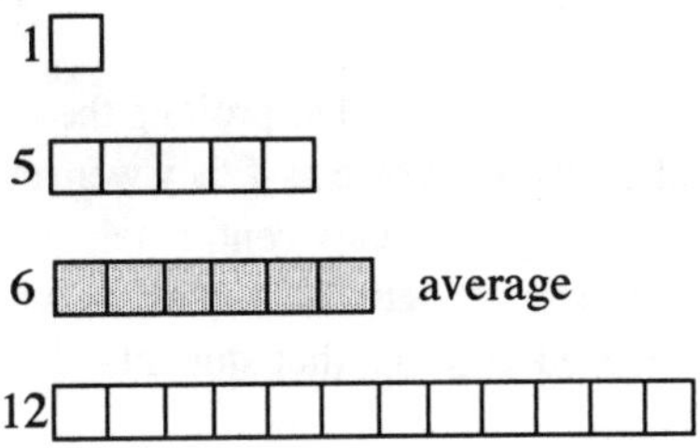

d) Write: 6 is the AVERAGE number of boxes.

e) Draw the average length.
It comes *between* the others.

f) Tom said the boxes were cans he found on the road. He finds an average of 6 cans a week.

2. To find the average:

   a) Draw 60, 40, 25, and 15 boxes.

   b) Guess the average number. Write it.

   c) Add all of the numbers.

   d) See 4 groups of boxes. Divide by 4.

   e) Write: Average = ____ boxes

   f) 60 is most. 15 is least.
   The *average number* is between. Draw it.

3. Draw to show these. Find the average number.

   a) 80, 72, 15 b) 100, 85, 15, 2 c) 17, 95, 40, 20

4. Find the average number of books: 20 books, 38 books, 150 books.

5. What do you say about an average of 6 candy bars a week for students in school? Write 5 ideas.

6. What helps you figure out the average of two numbers?

Drawing

Drawings tend to prolong the time students think about concepts and parts of drawings evolve as if they were events in a short story or landmarks on a map. The length of many centimeters becomes a decimeter and the average of three numbers fits between sequenced lengths. The development of "events" and the focused thinking that students do promote recall and new connections.

Drawings have other advantages. Students who compare denominators symbolically in *3/4 = x/8* see *4 less than 8* and infer that *3 is less than x*. They miss an opportunity, however, to connect the number and the sizes of eighths and fourths suggested by spatial representations (Figure 2).

Figure 2 Fourths and eighths

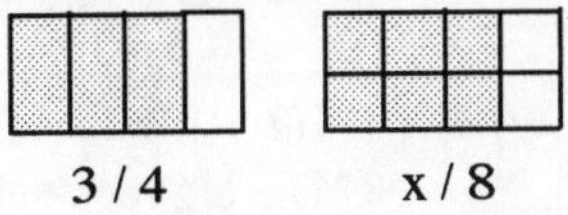

Astute mathematicians use drawings to check intuition and automatic response. Without a check, students commonly interpret 2/9 as a quantity greater than 1/3 because the terms of the fraction are greater. A drawing shows something else.

Figure 3 1 / 3 and 2 / 9

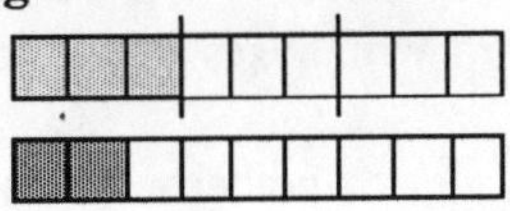

One way to increase connections between concepts and prior knowledge is *to listen to notation* and then make the words visible. A group of college students was struggling to illustrate 4/3 as a means of producing an equivalent expression, 1 1/3. An instructor asked them to read 4/3 aloud slowly. Students responded audibly: "Four-thirds." "Draw a third." "Four of what?" "Four of the thirds." After a few listening efforts, one of the students produced the required drawing. Searching their own words for connections evoked original thinking rather than direct recall, and understanding instead of algorithms.

Figure 4 4 / 3 = 1 1 / 3

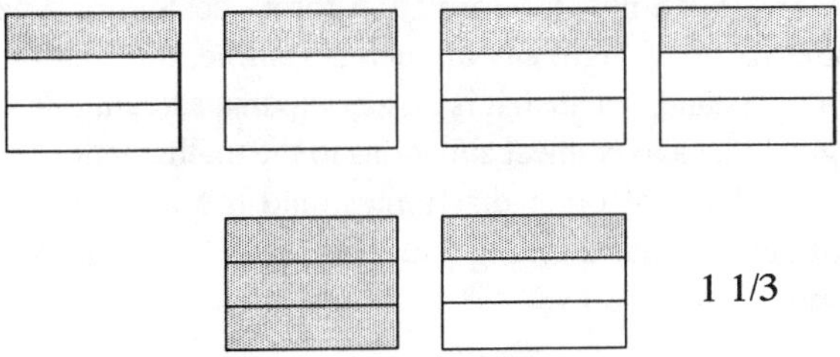

Teachers combine self-talk with drawings to communicate the effects of problem solving strategies. Students do it to evaluate their ideas. Working with fractional parts, they face the problem of assigning a name to the sum of unequal regions.

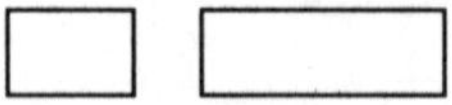

"1 + 1 = 2 unequal rectangles."

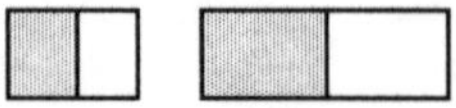

"1/2 + 1/2 does not equal 1."
"The big rectangles need to be the same size."

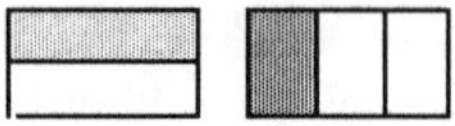

1 / 2 1 / 3

"How can I add 1/3 (area) to 1/2 in the other rectangle? 1/2 + 1/3 of what?"

"1/2 of the rectangle."

"Idea: Show the shaded parts with boxes all the same size."

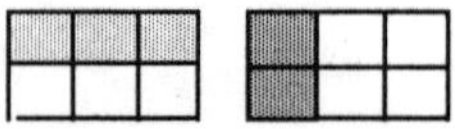

1 / 2 1 / 3

"Another way to show 1/2 and 1/3."

"Part of a rectangle and another part."

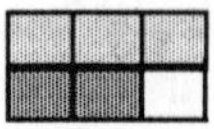

5 / 6

"Now it's easy to add."

Original Thinking

Original thinking, as it is used here, is the production of any idea new to a problem solver. It is a new interpretation for a question or a new solution. Although other people might say an idea is routine, the first time a student produces or understands an idea, it is *original* to that student. Problem solvers produce original ideas to connect situations to the mathematics they know, the knowledge they have of other disciplines, and life experiences outside of school. Without original thinking, students do not solve problems that are complex or novel.

Instructed to cut fifteen rectangles in half, each in a different way, students begin with the same four drawings.

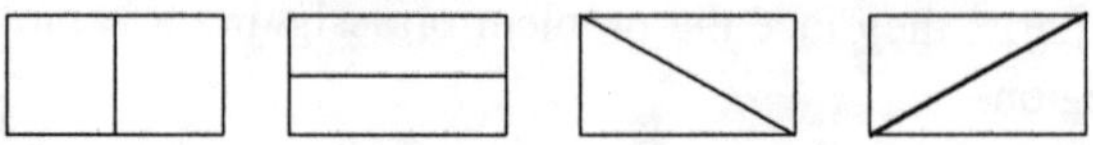

When short-term recall has no more to offer, original thinking creates a fifth drawing. At this point, some adults put their pencils down, assuming that no other drawings are possible. Others reveal a distorted sense of *half* by showing two parts that are not equivalent. Most are adventuresome, however, and continue to explore the problem by thinking beyond immediate recall. Some advance to a sense of *half* that includes disconnected parts and patterns.

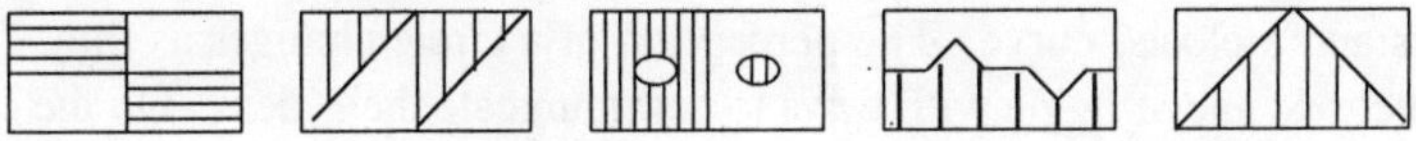

Thc purpose for the activity is to convince students that they are able to connect a task to experiences in other contexts and create possibilities that come from their own ingenuity.

In a fourth grade class, a team of students debated whether 34 divided by 2 was 17. The question was settled for one student by an announcement that 2 x 17 was 34. She captured an important aspect of problem solving when she sensed an agreement among given facts, an answer, and prior knowledge. Her tone of voice suggested that the idea occurred to her at that moment. Whether or not a teacher had emphasized the relationship between multiplication and division facts in a prior lesson, she produced the connection when it was helpful. Knowing that she is able to produce ideas should encourage her to experiment with original thinking in the future.

Without mentioning the word, *circle,* it is instructive to ask students to follow a few simple directions to create a drawing.

1. Mark a point on graph paper. Call it P.
2. Mark a point three units from P.
3. Mark many other points three units from P.
4. Mark all points on the paper three units from P.

Some students doing the activity for the first time do not consider points that lie outside the intersections of lines.

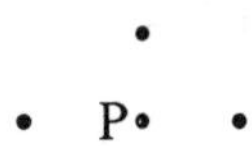

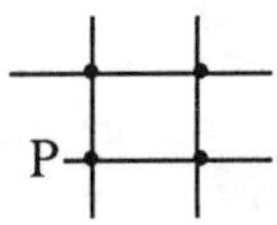

Marking a point inside a square takes *courage*. It is an assertive move characteristic of problem solvers and uncharactistic of students who rely heavily on immediate recall.

A discussion about *all* the points three units from P on the paper leads to new observations about units, distance, right triangles, and eventually to the nature of a circle. To understand circles, students integrate equal radii, *all* points and a closed curve. The perception of a circle changes as they make connections and struggle with ways to communicate their ideas. On the other hand, when the word, *circle*, is introduced in the lesson before the drawing, students think of a curved line without connecting it to all points a given distance from a particular point. To press for original thinking often means to make simple connections that would not occur without idea production.

Reading $3x = 7$ as a sentence, students look for a value of x that makes the statement true. Without a routine algorithm, they should appeal to their own sense-making. Productive thinkers look beyond routine resonses such as *x is 2 with 1 left over* and produce original ideas. "*X* is less than 7." "3 times something is 7." "3 groups of 2 things." "3 of more than 2 things." "How to get one *x* instead of three *x*'s? " "If 3 x is 7, then 1/3 of 3x and 1/3 of 7 gives x". "Groups of 3 and 3 equivalent groups. It is 2 1/3 groups of 3." One idea leads to another when students believe it will and when they are willing to identify and examine uncertain or faulty ideas as subproblems.

STRETCHING POWER

Self-directed problem solving is the mega-strategy, mathematical power through which learning merges with traditional notions of finding a solution in the face of a dilemma. At all levels of self-direction, students challenge themselves to build a tolerance for uneasiness that paves the way for more effective strategies. Another challenge is to develop monitoring that tests strategies and signals a need for new ones.

A program developing self-direction presses for flexible perspectives about learning and problem solving. With experience, problem solvers differentiate between ideas recalled directly from instruction and the constructive thinking they produce themselves. After simple recall, they look for applications and unstated assumptions, and persistence turns apparently unrelated ideas into a new point of view. The act of producing connections, however remote, prolongs the search for the best ideas. The most useful ideas represent new understanding or inventions that resolve conflict and build higher-order theories. The greater the challenge, the more students must stretch their thinking.

To make students aware of their choices, teachers ask for written descriptions of what students do in challenge situations. How do they begin to solve a problem? Do they expect to find answers immediately or direct idea production in many steps of a process? Do they persist and evaluate their efforts? When students sense that *I don't know what to do*, they consult their own list of problem-solving strategies.

Notes

1. Frank Smith, *To Think* (New York: Teachers College Press, 1989).

2. Robert Ennis, "Rational Thinking and Educational Practice" in *Philosophy and Education*, vol. 1, eightieth yearbook, ed. Jonas Soltis (Chicago: National Society for the Study of Education, 1981), 143-183.

3. Alan Schoenfeld, ed., *Cognitive Science and Mathematics Education* (Hillsdale, New Jersey: Lawrence Erlbaum, 1987).

4. John S. Brown, Allan Collins and Paul Duguid, "Situated Cognition and the Culture of Learning," *Educational Researcher* 18, no. 1 (1989): 32-42.

5. Janice Dole et al., "Moving from the Old to the New: Research on Reading Comprehension Instruction." *Review of Educational Research* 61, no. 2 (1991): 239-264.

6. M. Ann Dirkes, *Producing Ideas in Mathematics*, (Buffalo: United/D.O.K. Publishers, in press).

4 *Idea Listing: A Strategy*

Sensing ambiguity, problem solvers use many ways to bring the knowledge they have to consciousness and use this knowledge to develop structures.[1] With experience, the involvement compels them to explore the full extent of their mathematical power. Accepting the charge, students make significant learning and problem solving nearly synonymous.

To become problem solvers, students first need to learn how to produce a variety of plausible ideas.[2] That is, they learn to elaborate extensively before attempting to apply heuristics to a problem.[3] The production of multiple ideas helps them find the relevance of concepts to a given context[4,5] and construct possible analogies and models that support the spontaneous transfer of knowledge to ill-structured situations.[6]

In the ***SDM Program***, students produce ideas during informal interaction and in idea-listing sessions that have more structure. The knowledge goal is the same for both formats, namely, to activate what students know and to connect prior knowledge to new concepts. Students develop a sense that *idea production depends on me.*[7] They assume responsibility for constructing mathematical concepts[8] and integrating ideas internally. Connections between the number system and the metric system arise, for example, when students produce what they know about both systems and make conjectures about their relationships. In the initial learning stages, students need reminders to produce ideas and seek connections. Later, they use many strategies without conscious direction.

INFORMAL AND FORMAL PRODUCTION

Because idea listing is a way to begin thinking, students use it to respond to dead ends and feelings of insecurity. Marie, a child in the second grade, asked for help to complete the statement, $? + 2 = 5$. The teacher suggested that she say everything that she knew about it. Marie's response, listed in Figure 1, refuted the child's claim that she didn't know what to do.

Figure 1 Ideas on *? + 2 = 5*

I see 2 and 5.	It's like putting blocks together.
5 is bigger.	3 + 2 = 5
= means equals.	2 + 3 = 5
+ means plus.	If = was +, then 7.
0 0 + 0 0 0 = 5	What and 2 is 5?

In a similar response to *I don't know*, a teacher asked Carlos, a student in the fifth grade, to list guesses for writing *three-hundred-five thousandths* in numbers.

Teacher Write some guesses.
Carlos 305.

Teacher Write another guess.
Carlos .305

Teacher Write another guess.
Carlos 30.5

Teacher Now select the best guess.
Carlos Points to .305

Teacher Who wrote all of the guesses and picked out the best one?
Carlos I did. Thank you for helping me.

The teacher's *help* was asking Carlos to produce possibilities and making him aware of his own ability to produce ideas. The process introduced him to writing decimal expressions, a process he was now ready to analyze.

Idea listing is a more formal strategy used to cultivate divergent thinking, sustained effort, and confidence in the construction of ideas. The ideas produced are inventions or elaborations on what students know and what they believe. Some are recalled from long-term memory and others more recently from teachers and textbooks. To construct understanding, problem solvers stretch their knowledge into new combinations and associations. Some of these ideas are formed by placing known concepts in unfamiliar contexts and unexpected ideas in familiar situations. Lists are likely to include the best connection or solution that a student can make because they represent prolonged thinking and active interaction with mathematics.

Student lists written about a concept or in response to a word problem provide opportunities to study the knowledge accessible to students and the risks they will take to make connections. The length and quality of a list suggests prescriptions. Students who produce abundant ideas have a better chance for seeing connections[9] and generating new ones than those who list only a few ideas. Long lists usually show that students have considered a wide range of possibilities for transfer to new contexts. After many ideas are in view, students select the best ideas and organize them in a problem solution or the overview of a concept. Comparisons lead to self-correction and invention.

As admirable as mental arithmetic is, few students master it and the immediacy of the process misleads many into thinking that they should do complex problem solving quickly. "Found a quarter, lost a nickel, received a dime from my brother." Individuals make the decision to write and explore the situation or accept the challenge to reason without the aid of a visual presentation. If a mental conversion to 25 - 5 + 10 is unsuccessful, students write possibilities for the expression rather than expect to do the computation as quickly as the data is given orally. Sensing difficulty, they list ideas to keep track of complex data and compare it to solutions.

DISPOSITIONS AND TECHNIQUES

A student studying to be a teacher explained that idea listing helped her solve problems because it helped her think. After struggling with the strategy for a few weeks she wrote in a journal:

> Listing ideas before solving a problem is very hard for me to do. I have always been taught to give an answer right away and not dwell on it too hard. I would like to work on listing ideas about my problems as they come up. In order to teach children to take time and think about a problem, I must learn to think for myself first.

Asked to say *everything you know about numbers*, children in the first grade gave responses that a teacher wrote on the chalkboard. The sequence of their ideas was, *1, 2, 3, 4, 5, 6, 7, 8, 100, 10, . . .* The child who said *100* broke the memorized pattern. At age six, she was thinking for herself. The response gave the others in the group a lesson about exploring categories of ideas. Soon there were diverse ideas like *425 on my house* and *1/2 an apple*.

Although students with a strong mathematical background produce ideas more easily than those with more limited knowledge, everyone can generate observations and examples from their own experience. The assumption is that everyone can think, and that the quality of a list depends on dispositions for thinking as well as knowledge. Curiosity and commitment, for example, account for the time students are willing to allocate for thinking. Individuals choose either impulsivity and premature closure[10] or persistence that sustains productive thinking. A willingness to think probably influences productivity more than minor variations in age or prior achievement.

Other dispositions that support problem solving include a tolerance for ambiguity during the consideration of contradictory ideas. Problem solvers actually create complexity in the production of new ideas for solutions or deeper understanding of concepts. Expecting to grapple with uncertainty and incomplete knowledge, they value the disorganized part of problem solving as positively as the representation of an orderly solution.

At the end of one listing session, a group leader asked participants to describe their goals for the session. "What were you trying to do?" Some students said they focused on recall or worked for a large number of ideas. One said she wanted to find out what she could do, the capacity of her brain. Others were not yet conscious of a particular focus.

Asked about the effect of idea listing, many revealed a readiness to think independently. Constructive thinkers said, "I didn't know I knew that much." "I like to say what I think." "I knew A and I knew B, but I never thought of saying how A and B explain one another." Those who are less productive indicated, "It's hard work." "I was going to give other ideas, but I thought they were wrong." "What did you want me to write?" "When you said all ideas were acceptable, I thought you were trying to trick us." "I like multiple-choice answers better." In the beginning, students are more likely to consider idea listing as an isolated exercise rather than an aid to finding problems and solutions. With experience, they monitor strategies because they value their own thinking and ways to direct it toward an objective. Instead of asking whether they should do *this* or *that*, students learn to make the decision themsleves. Often the best response is to do both activities or list both ideas.

To develop confidence in idea listing, students begin with an everyday topic. *Naming birds, for* example, most people find they can list more ideas than they anticipate. (See Chapter 2). The intent is to transfer this assessment to mathematical concepts and word problems. Ultimately, the distinction between the kinds of thinking used in mathematical and nonmathematical problems is blurred.

Topics Used to Focus Idea Listing

1. Everyday thinking: foods, cars, birds

2. Mathematical concepts: fractions, geometry, graphing

3. Word problems: nonmathematical social problems; mathematical problems generated by students, prescribed problems, insight problems

Questions with more than one answer help students break away from rigid expectations for convergent responses. Topics from other disciplines and everyday situations are included to suggest that idea production is a thinking strategy and a habit that can be cultivated throughout the day. Until students associate fluency with idea listing, it is advisable to stipulate the number of ideas required. For some students, "List many ideas" infers two or three ideas.

Questions with Many Answers

1. List five questions about a newspaper article.
2. List five ways to write 1,286 or $3 + 6 + 4 = 13$.
   Write ten ways to keep your body healthy.
3. List five reasons for keeping your papers in a folder.
4. Choose a noun. List everything you know about it.
   List many ideas about *problem solving* or *learning*.
5. List six differences between a city and a town.
6. List four reasons why $4 + 19 = 23$.

7. Draw five large rectangles all the same size.
   Cut Rectangle A into 4 equal parts.
   Cut Rectangle B into 4 equal parts in another way.
   Cut Rectangle C into 4 unequal parts.
   Cut Rectangle D into 8 equal parts.
   Cut Rectangle E into 6 equal parts.

   Shade one part in each rectangle.

   You want to put two parts into the same rectangle.
   Which pair of rectangles makes this possible?
   Another pair? Why?

Very long lists demand flexible and original thinking because ideas common to one category and to a large number of people appear in only a few entries. Jana, a college student, compiled this list to stimulate abundant thinking.

Jana's Techniques

1. In the beginning of list making on a concept, write fast so that I will not miss any ideas from the flood coming.

2. First number the paper up to a large number (even 1 to 100); then try to fill the spaces. It's too easy for me to stop at 5 or 10 items if I number as I go. Do not worry about wasting paper. Have plenty. The answer could lie hidden, buried without this extensive writing.

3. Clear my concentration. If there's something bugging me while I'm making a list, take care of it first. Write a note if necessary. Then produce more ideas with a clear mind.

4. Write messy if I want to. It's my list. Print, write, doodle. It's okay. I'm thinking.

Developed originally for language arts, reciprocal teaching also stimulates idea production. Students assume the role of teacher in small groups to elicit questions, clarifications, a summary, and predictions about text material.[11]

Listing ideas on a mathematical concept in a group serves a similar purpose. Whether students are interpreting a narrative section or a mathematical concept, group interaction helps them probe, connect, and reflect. The objective is to generate ideas[12] and construct meaning.[13] That is, the socialization process increases understanding and individual thinking.[14,15]

In a group, students encourage one another to think and suggest new trains of thought[16] even after estimating that they have no more ideas to offer. Ideas produced by other group members become a springboard to more complex thinking, thereby having the effect of increasing the potential of listeners. The interaction prepares students to participate in consensus or at least in negotiated discourse that respects diverse ideas expressed through informal language and different types of representations.[17]

PROCEDURE

An introduction to idea-listing includes a statement of purpose, namely, to generate abundant ideas about a concept. Students follow these guidelines and seek many interpretations for them:

- List as many ideas as you can.
- Avoid evaluating ideas during the listing.
- Include many categories.
- List unusual ideas, some that are "far-out."

After discussing self-imposed attitudes, many groups use positive self-talk to stimulate thinking.

> "I will say what I know about it."
> "I know more than I think I do."
> "I will think longer than I think I can."
> "I will produce ideas that no one told me."
> "I will produce many different kinds of ideas."

Next they examine the validity of negative self-talk.

> "I don't have ideas. I'm not creative."
> "Other people laugh at my ideas."

"What I say might be wrong."
"I never did get math."

It is important to defer judgment until many ideas are in view so that evaluation does not curtail thinking[18] except to make entries respond generally to the given topic. Participants understand that ideas are possibilities and a stimulus for more ideas rather than expected certainties or finished products. They are willing to reconsider ideas that at first appear illogical or irrelevant as precursors to invention and insight. When productive efforts are exhausted, lists are evaluated according to selected criteria, e.g., improvement in the number and variety of original ideas, mathematical understanding and sophistication, and probable transfer to problem solving.

Listing sessions take place in large groups, small groups, and individually. Group size or a combination of group and individual list-writing depends on the pedagogical purpose for a session. With only one person in a group writing ideas, others can concentrate entirely on idea production. For large groups, a leader writes ideas on a chalkboard or flip chart without comment. In small groups students can see what one person writes on paper. Some people claim, however, that the act of writing ideas themselves facilitates thinking as it does in journal writing.

A leader announces the topic and signals the time to begin. Constructive thinkers let one idea lead to another freely. Searching for new insights, they expect to fumble and only gradually clarify the concept beyond initial perceptions. The more extensive the list the greater the assurance that diverse elements have been considered for the construction of new ideas and that internal resources have been sampled adequately.

For mathematical listings, a checklist of several categories helps students focus on appropriate concerns.

Idea Listing Checklist

definitions	spatial observations	reasons why
drawings	properties	uses
statements of equality	patterns	questions
estimates, inequalities	meanings	

Most lists are enriched further by representations in words, drawings, and symbols and a distinction between naming vocabulary words and explanation or application. A decision to cite specialized categories depends on the flexibility expected. The more categories included in a checklist, the fewer there are for students to generate in their lists. An introduction to these categories, however, is instructive. Some of these are length, area, volume; whole numbers, decimal parts; grouping and renaming. Drawings are used to illustrate both given data with the answer and life contexts for it. Students deal with a few categories at a time, gradually increasing the number to enlarge their perceptions of mathematical thinking. The ultimate goal is to make their lists reflect many categories naturally without referring to a checklist.

At first, students count all of the ideas on a list to find a score. Later, the score is the number of mathematical categories. A liberal scoring of subcategories and other variations encourages flexible thinking that leads to new understanding. Points can be doubled for unique ideas and subtracted for the omission of a basic concept emphasized during instruction.

EXAMPLES

In the beginning, most of the entries on student lists are factual in nature and fragmented bits of knowledge. There is a tendency to begin lists with mathematical vocabulary and formulas in textbook fashion. Simple examples from everyday experience might come next. Ideas that are the most fundamental to a concept or unique within a group often appear near the ends of lists. Students who list only a few ideas either fail to cite important concepts or list them immediately, demonstrating that they did not need to do idea listing for a simple task or they chose not to explore possible implications.

Mundane observations often familiarize students with a topic and suggest better ideas. If students list only ideas with a high degree of relevance, the degree of problem solving is minimized. Some would say that *they do not have a problem.* A large number of ideas usually indicates a search that will deal with complex data (complex for the problem solver).

At advanced levels of idea listing, students learn to convert unconventional or "strange" ideas to useful ones and make connections among ideas that represent new knowledge. Category checklists and brainstorming techniques[19] help them search for relationships they know but seldom use. Some students say they gain insight by simply combining ideas in ways that had not occurred to them previously.

With only a few prior class sessions on idea listing, five students in the fifth grade produced ideas on "division by four." Numbers in parentheses (Figure 2) indicate categories and a point count. Some of the categories used here are vocabulary, computation, estimation, drawing, and applications.

Figure 2 Division by 4

(1) division
divisor
quotient
remainder
(2) $12 \div 4 = 3$
$36 \div 4 = 9$
$32 \div 4 = 8$
$28 \div 4 = 7$
$24 \div 4 = 6$
$20 \div 4 = 5$
$16 \div 4 = 4$
$8 \div 4 = 2$
$4 \div 4 = 0$
(3) math
(4) even numbers
(5) less than division by 5
more than division by 3
(6) separate
(7) see how many fours in a number
(8) 2 goes into 4 two times
(9) divide 4 into large number
(10) 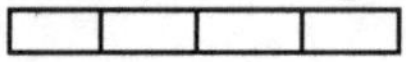 in four parts
(11) fractions
(12) four pieces of pie
(13) 4 quarters in a dollar
\$1.00 4 of 25¢
(14) 40¢ 4 of 10¢
dividing into parts
4 parts

In this list (Figure 2), students did not associate *division by 4* with fraction notation as in *12/4* or with remainders in *13 divided by 4 is 3 groups and 1/4 of another group of 4*. Notice that the initial response was vocabulary and other memorized facts. Listing these ideas is almost a prerequisite that students impose upon themselves. When idea listers tire of ordinary ideas, they turn to new categories and original thinking. A study of hundreds of lists produced by students in many age groups suggests that the first ideas produced are those that receive the most attention in school. Ideas at the ends of long lists come from probing and imagination.

If students omit ideas central to a concept, it is doubtful they would use them to solve word problems. It is common for students to identify one interpretation of division and omit the other. Some think of *four equivalent parts*, and others *groups of four*. Many students, adults included, have never learned a second interpretation or they do not have ready access to it.

A subsequent discussion might deal with interpretations of division and with the magnitudes of quotients obtained from dividing the same number by 3 and 5 (Item 5, Figure 2). Number patterns and a comparison of dividing objects (number), length, area, and volume would also be appropriate.

Entries in Figure 3 were compiled from lists generated individually by twenty-eight college students after practice with idea listing on other topics during two class periods and a written assignment. Students were enrolled in a mathematics methods course for elementary school teachers.

Figure 3 Idea Listing on Fractions

parts, pieces, whole
numerator, denominator
2/ 3 and 175/3
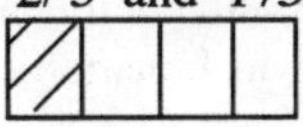
1 / 4
13 / 13 = 1
13 / 14 < 1 14 / 13 > 1
fraction bars, C-rods
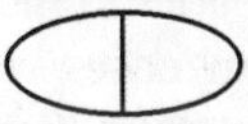
1 / 2
$.25, $.75, quarters
1/2 is groups of 2
8 / 3 is 8 divided by 3
1 / 4 over 4 / 5
music: 1/8, whole notes
recipes, shoe size
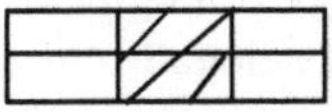
simplify
1/3 = 2/6
"Broken" numbers
cut equal parts

one number over another
1/3 + 2/3 = 3/3
2/3 − 1/3 = 1
Hard for little kids
Divide bottom number into top one.
3/4 + 75%
Not used on computers
equivalent fractions
mixed numbers, 2 1/4
answer in division
pizza
pie graph: parts income
can be percentages
telling time, half-hour
The larger the bottom number, the the more the pieces.
7/8 is almost 1
lowest terms 6/8 = 3/4
common denominator to add
anything over anything = 1

In lists about fractions, some students write *less than one*, without recognizing fractions *greater than one.* This restriction could limit estimations for sums that exceed *one.*

Figure 4 shows ideas from individually generated lists that deal most directly with meaning. Although Student C probably applies her knowledge to problems best, the most desirable idea list would include responses from all three students.

Figure 4 Ideas about area

Student A Area is the inside flatness of a rectangle.
Area is all of the insides of squares.

Student B 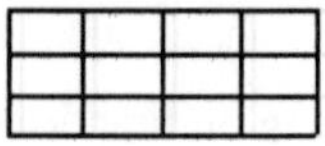 A = l x w A = 12

Student C 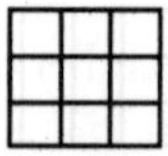  1 cm 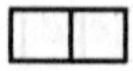

1 cm 2 cm²

The interior of
9 squares

Long lists tend to correct hasty responses. Student B, for example, might have looked over the list in the hope that an items would suggest something else. The new idea could be another drawing of twelve sections to show area. The drawing might show squares or labels indicating square inches. To make the activity even more beneficial, students identify the source of their ideas. They made squares out of habit, or they made a connection between the familiar notion of *square inches* and a *drawing with squares.*

Four lists written individually by adults, show much variance in mathematical knowledge and invention. The adults were not enrolled in school at the time and had not studied techniques for idea listing. In Figure 5, a single star designates the more useful mathematical ideas and a double-star concepts that require clarification. Notice that the writer of List B ignored decimal points and the multiplication operation in *.9 x .9 =* .

Figure 5 Idea lists on .9 x .9 =

List A

.81
multiplying
equation
use decimals
can change to %
has numbers
has equals sign
math problem
answer contains 2 numbers
*doesn't equal 1.00
has 4 numbers totally
has only 9, 8, 1
3 different numbers
*doesn't equal a whole number
*can be converted to fraction
isn't adding, subtracting
written in pencil

List B

.81
4 x 9 = 36 inches
3 x 9 = 27
81 days is 2 1/2 months
$81 is $19 from $100.00
81 days until summer I hope
81 inches of snow. Oh, now.
9 beautiful grandchildren
9 room house with maid
9 days pay, 9 gifts
9 yards material
9 sq. yds. carpet
81 piece setting of dishes
81 different desserts, T-bones
9 pieces of new furniture
81 people in a wedding party
9 days to get the house clean
81 minutes to a soap opera

List C

.81
measure
more than one figure
*% of one
IQ
numbers, learning
some of two numbers
alpha number with numbers
test scores, testing
used by accountants
multiplication
answer to a question
** number times another number
totals more than either of
the numbers themselves
*period before, never a whole nb.
multiply only

List D

.81, two nines, produces 8 and 1
*square root of .6561
computation, multiplication
equation
decimal
*numeric symbols
*$.9^2$, $.3^4$, 9^{-2}, 3^{-4} **
a low grade point average
rational number, fraction
real number
**inverse of 10,000 / 81
*probability of 2 events of
.9 probability each
*> 80%, < 100%, > 4 / 5, almost 1.0
summation sign from 1 to 2
.9 (with x subscript)
**repetitive addition
equality function, question

Students generally like the freedom of idea listing without evaluation, but soon feel the strain of increased expectations for extended thinking and new connections among ideas. Many choose to produce ideas freely before referring to a category checklist, and then challenge themselves to exceed the average number of ideas in their lists. They are quick to acknowledge that they were filled with uncertainty as they wrote the first few lists they attempted alone. A sense of accomplishment increases, however, as students learn to use lists for inventing knowledge and identifying misconceptions.

On the other hand, some students observe that they have used variations of idea listing as a thinking strategy for years. Variations include listing many possibilities for spelling a word before selecting the most likely one and making oral attempts to recall or pronounce a person's name "on the tip of my tongue." Some students list advantages and disadvantages to help them make social decisions. Idea listing sessions give students a name for what they do, bring fluent productive thinking to mathematics, and offer techniques for making it effective.

Certainly, older students should produce ideas in many categories and more abundantly than younger ones. Although older students draw from richer life experiences and more mathematics courses, very young children are not hampered as much by expectations for established mathematical procedure. For everyone, thinking requires focus, planning, and concern for *what I think.* Individuals need to assess their willingness to respond to the challenge of idea production. Questions with many answers, a procedure for listing ideas in groups, and a mathematical checklist are strategies that develop confidence in the process.

Notes

1. John. D. Bransford, C. K. Kinzer, and V. Risko, "Teaching Thinking and Content Knowledge: Toward and Integrated Approach" in *Dimensions of Thinking and Cognitive Instruction*, ed. Beau F. Jones and Lorna Idol (Hillsdale, New Jersey: Lawrence Erlbaum, 1990).

2. Alan H. Schoenfeld, "Measures of Problem-Solving Performance and of Problem-Solving Instruction," *Journal for Research in Mathematics Education*, 13, no. 1 (1982): 31-49.

3. Richard E. Mayer, "Implications of Cognitive Psychology for Instruction in Mathematical Problem Solving," in *Teaching and Learning Mathematical Problem Solving*, ed. Edward A. Silver (Hillsdale, New Jersey: Lawrence Erlbaum, 1985), 123-138.

4. Ann L. Brown, John Bransford, R. Ferrara, and Joseph Campione, "Learning, Remembering, and Understanding" in *Carmichael's Handbook of Child Psychology,* vol. 1, ed. John Flavell and E. M. Markman (New York: John Wiley Publisher, 1983), 77-166.

5. Herbert A. Simon, "Problem Solving and Education" in *Problem Solving and Education: Issues in Teaching*, ed. D.T. Tuma and F. Reif (Hillsdale, New Jersey: Lawrence Erlbaum, 1980), 81-96.

6. John Clement, "Learning Via Model Construction and Criticism: Protcol Evidence on Sources of Creativity in Science" in *Handbook of Creativity*, ed. J.A. Glover, R.R. Ronning and C.R. Reynolds (New York: Plenum Press, 1989), 341-382.

7. Larry L. Hatfield, "Preparing Early Childhood Teachers for Constructive Mathematical Environments," in *Transforming Children's Mathematics Education,* ed. Leslie P. Steffe and Terry Wood (Hillsdale, New Jersey: Lawrence Erlbaum, 1990), 407-414.

8. National Council of Teachers of Mathematics, *Curriculum and Evaluation Standards for School Mathematics* (Reston, Virginia: NCTM, 1989).

9. Mark Johnson, *The Body in the Mind* (Chicago: The University of Chicago Press, 1987).

10. Frank K. Lester Jr., "Methodological Considerations in Research on Mathematical Problem-Solving Instruction" in *Teaching and Learning Mathematical Problem Solving,* ed. Edward A. Silver (Hillsdale, New Jersey: Lawrence Erlbaum, 1985), 41-69.

11. Ann L. Brown and Annmarie Palincsar, "Guided, Cooperative Learning and Individual Knowledge Acquisition" in *Knowledge, Learning and Instruciton*, ed. Lauren B. Resnick (Hillsdale, New Jersey: Lawrence Erlbaum,

1989), 393-451.

12. Merlin C. Wittrock and Kathryn Alesandrini, "Generation of Summaries and Analogies and Analytic and Holistic Abilities," *American Educational Research Journal* 27, no. 3 (Fall 1990): 489-502.

13. Lauren B. Resnick, "Treating Mathematics as an Ill-Structured Discipline" in *The Teaching and Assessing of Mathematical Problem Solving*, vol. 3, ed. Randall I. Charles and Edward A. Silver (Hillsdale, New Jersey: Lawrence Erlbaum, 1989), 32-60.

14. Lev S. Vygotsky, *Mind in Society* (Cambridge, Massachusetts: Harvard University Press, 1978).

15. Jane Healy, *Endangered Minds* (New York: Simon and Schuster Publishers, 1990).

16. Arthur Whimbey and Jack Lochhead, *Problem Solving and Comprehension* (Hillsdale, New Jersey: Lawrence Erlbaum, 1982).

17. Vicky L. Kouba, "The Making of Meaning: A Model for Interpreting Mathematics Education Research," paper presented at the annual meeting of the American Educational Research Association, Chicago, 1991.

18. Magdalene Lampert, "When the Problem Is Not the Question and the Solution Is Not the Answer: Mathematical Knowing and Teaching," *American Educational Research Journal*, 27, no. 1 (1990): 29-63.

19. Alex Osborn, *Applied Imagination* (New York: Scribner and Sons, 1963).

5 *Word Problems and Other Problem Solving*

Students usually think about problem solving as a process used on scenarios described in words. Their job is to select the mathematical operation that produces an expected answer. The notion of problem solving expands when students see themselves as idea producers for any situation that poses a dilemma, whether the uncertainty appears in words, symbols, drawings, or people. Listing ideas about concepts begins the process. If students are the ones to state facts and conditions that agree with their own interpretations of life, they work in an open situation. If someone else wrote the problem, as in a closed situation, they focus on the interpretation intended by the author. In each case, solutions and learning depend on prerequisite knowledge and the construction of new ideas.

Several conditions pave the way to understanding the nature of problem solving: (a) encountering novel types of problems continuously, (b) encouragement to seek comprehension over efficiency, (c) freedom from an urgent need to get external reinforcement, and (d) dialogical interaction.[1] Under these conditions, problem solvers are willing to think without having a clear direction. They develop much more confidence and self-direction than those who depend largely on routine thinking.

To accept the challenge of problem solving, students must recognize further that the search for comprehension and new ideas will include error and false leads. Problem solvers examine these digressions for misconceptions and clues to solutions. They look at errors as a prelude to better ideas. Even random computation can contribute to a solution if students examine it for connections to given information and the objective for their work. Telling students not to produce error, on the other hand, often confirms a belief that *problem solving is not for me because I am supposed to know what to do immediately.* A likely inference is that routine skills are the only option.

Two girls in third grade were about to give up on a problem when a teacher encouraged them to write what they knew about symbolic data.

$$\square + 2 = 6$$
$$\diamond - \square = \square$$

Although these students had substituted *4* for the box in the first equation, they claimed that they could not write the next number sentence because they did not know what to write for the diamond. "Write what you *do know* in the sentence." After some coaxing to write whatever they were thinking, one student wrote $\diamond - 4 = 4$. At this point she recognized the solution. With more numbers in view, she could substitute *8* for the diamond.

Writing almost any number for the diamond extends thinking. Students can try $10 - 4 = 4$ and $3 - 4 = 4$, or make a drawing that shows taking away *4*. The major task for students is to write under the pressure that comes from uncertainty about the result. Once this obstacle is overcome, writing and reviewing ideas activates prior knowledge and new connections.

Some teachers object to the inefficiency of the process. Why clutter thinking with error? One answer to this question is that problem solving and learning require multiple connections to sort out complex ideas. For the person who says, "I don't know," inference is neither simple nor automatic.

The difference between routine thinking and idea production leads to a distinction between automatic answers and problem solving. When students face a dilemma, they must become idea producers. Teachers tell them, "Now you are thinking like a problem solver. The work that it takes is satisfying." Then students know that they should experiment with ideas. Some business consultants call this *fast failures* that contribute to learning and the production of solutions.[2] The argument does not reinforce error or support ignorance. Rather, it urges accountability for dispositions that support a continual search for better estimates and a recognition of contradictions.

LISTS FOR WORD PROBLEMS

Idea listing is a strategy that helps students work on a problem regardless of the perceived difficulty level. Rather than succumb to the urgency of producing a solution in a perplexing situation, students use idea listing to search for connections that *might* lead to a solution. Producing ideas with an *unschooled mind*[3] as well as an academic posture, students generate facts and possibilities to make sense of a dilemma. "What do I know about it?" "What might be connected to the situation?" With multiple answers to these questions, problem solvers are ready to organize a response to a problem.

In closed problems, idea listing reaches into the meaning of words and symbols. Restating the question in more than one way calls for the production of familiar language that shapes an understanding of the intended objective. At the same time, students monitor their progress. "Is this what I am to do?" "Will another statement tell me more clearly what I am to find?" To visualize a context, some students add conditions or facts that are probable but do not change the nature of the problem. Their own input, conscious or not, contributes to analysis and offers a direction for their work.

An array of techniques increases idea production within a general plan for solving problems. Techniques were introduced in Chapter 4 and the plan is developed in Chapters 5 and 6.

Techniques for Idea Production (TIP)

1. Distinguish between algorithms used in familiar situations and sustained idea production in situations that are complex or new.

2. Use idea listing with ground rules: defer judgment, aim for fluency, flexibility and originality. List ideas from practical experience, invention, and recall.

3. Use a checklist to direct idea production into mathematical thinking: estimates, equivalencies, definitions, observations, uses, special properties; drawings, symbols; words; whole numbers, fractions, and other notation.

4. Produce lists in small groups, alone, and with an entire class. One person records ideas on paper or a chalkboard. All group members contribute.

5. Score lists to monitor (a) the number of ideas produced and (b) the flexibility of these ideas (categories of ideas). Mark inventions and new insights.

6. Prompt yourself: "When in doubt, write it down." "All ideas are valuable, some more valuable than others." "I can decide what to do."

Problem Solving Plan for Word Problems

Facts List the given facts. List other probable facts and assumptions.

Question State the question; it tells what is to be done. Restate it in your own words in more than one way to help you understand the objective. Name the *thing* to be found, e.g., square inches, kilograms, apples.

Mark the best statement of the question.

Ideas List related mathematical ideas and possible leads to a solution. Emphasize ***TIP*** techniques here, especially drawings and number sentences.

Will the answer be *large*? About how much? Compare the meaning of operations (+, – , x, ÷) to the facts and question. Take time to think.

Mark the answer.

Check Look for agreement among the facts, question and answer.

Students should have ample experience with the format of the plan before complicating the process with a fine distinction among facts, questions, and ideas. Soon, however, teachers point out that computation trials appear in the idea section and *If . . .* clauses under facts. Consider the sentence, *If Lenny paid $2.00 for 3/4 pound of candy,* what would be the cost of a pound?" "*$2 paid for 3/4 pound* is a ***Fact***. *Find the cost of a pound* is the ***Question***. As analysis improves, problem solvers learn that attention to different kinds of information structures their thinking. The presentation of the problem becomes a work of art, a synthesis of elaborated facts, a question, and possible solutions.

Working with a partner increases idea production and helps students keep track of the steps in the plan. One student asks questions from a prepared list and writes whatever another student says without offering suggestions or solutions. The second student answers questions to solve the problem.

Questions for a Problem Solver

Challenge

1. Read the problem. Do you recognize what to do right away?
2. How much data is here? A little or very much?
3. How difficult might the problem be for you?
4. How much time do you need for it?

Facts

5. What are the given facts? Put *if* clauses here.
6. What else can you say about the facts?
7. What else is probably happening?
8. Did you miss any ideas about quantity?

Question

9. There is something that you do not know.
   What are you to find out?
10. Say it in your own words. Another way.
    What does it mean?
11. Which statement of the question describes the question best?
    This is your objective. I will star it.
12. What else do you need to know?

Ideas

13. What will the units be in the answer? Miles? Square feet?
    Hours? Number of things?
14. Will the answer be larger than ______?
15. Choose an operation. Add, subtract, multiply, divide? Use it.
16. Is there more than one step? Will you have to find more
    than one answer? Do it. What else?
17. Read the facts again. Read the question that is starred.
18. What else can you try?
19. How shall I label these numbers? Is this length? Hours?
20. Which computation says how to solve the problem?
    What else? Is this enough? Label **Answer**.

Check

21. Does the answer really answer the question?
22. Does the answer make sense? Too big?

Monitoring
23. How much of a challenge was this problem for you?
24. Did you stretch for ideas?
25. How satisfied are you with the solution?

APPLICATIONS

Where persistence and concern for the artistry of problem solving are absent, students are likely to claim that a general plan with many steps "is too much trouble." This assessment is correct for simple, closed mathematical situations. If a situation is a challenge and intuition fails, however, an organized approach can help. The difficulty of a problem and the willingness of students to think each call for decisions about whether to use the plan and, if it is used, the degree of idea listing appropriate for each step.

Two examples illustrate these choices.

Bus Driver Situation A bus driver drove 45 kilometers each day on a school bus route. She drove the route twice each day, five days a week for 180 days each year. How far did she drive in a year?

Example 1 Routine thinking in a situation familiar to the student

Facts: what is known
bus and female driver
school bus route
45 km one way
45 km two times per day
5 days a week
180 days a year
Other facts added Driver stayed at school all day. The question asks only for distance traveled on the bus route, not to and from the driver's home.

Question: the objective or what to find
How far did the bus driver travel each year?

Ideas computation, strategies
45 km x 2 = 90 km / da

90 x 5 da = 450 km/wk (Digression)
180 da x 450 = ? 180 da x 90 = 16200 km / yr

Check Distance for 45 km one way, twice a day, 180 days / yr.
Score: 15

Example 2 shows how another student in the class worked on the same problem without finding an answer. This problem solver recorded facts from the narrative, but did not use enough self-help, e.g., take time to interpret facts, connect ideas to the question, and make drawings. He stopped thinking too soon and probably neglected to examine the ideas listed.

Example 2 The work of a student who did not know the mathematics to use on a given problem

Facts
45 kilometers each day
bus route
twice each day for one year
five days a week
180 days each year

Question
How far did the bus driver travel each year?
Where did the bus go?
How far did the bus driver travel on the school bus route each year?
Find the distance the driver traveled.
* Find the number of kilometers the driver traveled.

Ideas
Round trip 45 km x 2
5 x 2 is a week's km, like 10 days a week
90 x 10 = 900 km in a week
What else? Am I finished?
No, it is 45 x 10 = 450 km in a week, 180 days in a year
180 x 450 = Not sure.

Check I found the km using round trip 5 days in a week.
Score: 18

Students without a direct solution should expect to generate many more ideas than one who finds a solution easily. In unfamiliar situations they remind themselves to aim for a score of at least 25 to 30. For the most challenging problems, they might produce 50 ideas and plan to return to the problem the next day. For teachers, establishing this perception of problem solving is worth the effort. Without it, students decide to accept the results obtained without their full mathematical power. Effective problem solving requires that students control steps of the plan as well as the production of mathematical ideas.

Decisions extend also to perceptions of risk. A college student explained that problem solving through idea listing took courage.

> I didn't feel confident about listing ideas because it wasn't clear to me what was an acceptable answer and which ideas should be grouped together.

Describing her usual approach to mathematical problem solving, another college student expressed a similar concern.

> I didn't realize, until this course, how dependent I had become on the instructor to tell me what I had to do in a class or what was expected of me. I was very surprised at myself, . . . seeing that self-direction can bring about self-doubt. If I am faced with a problem that is not familiar, I tend to panic and feel as if I can't do it. My education focused on drill and practice, not self-direction.

After learning to produce ideas freely, this student determined to press for all that her brain could muster. On the final examination for that course, she demonstrated this resolve by using all the time allowed rather than follow her usual habit of finishing tests as soon as possible. Both students learned that risk and persistence were basic to problem solving.

EXPERIENCE WITH THE FORMAT

An introduction to the problem solving plan emphasizes time for learning how to make the steps useful on complex and novel situations. Only challenge problems help students recognize the value of listing facts, questions, and ideas to search and probe for new ideas. Closed situations that prolem solvers can handle through a simple or straightforward approach, however, make the advantages of the plan remote.

Most students need to work on the ***Fact-Question-Idea*** plan extensively before they will convert old habits into new ones. Responses from many groups suggest that a few class sessions on the plan are not enough for them to understand the purpose for each step and integrate idea production into the process. Problem solvers should recognize also that faulty efforts and resistance to difficult work are signs that a situation warrants the prolonged thinking demanded by the problem-solving plan or a variation of it.

Typical performance in a seventh grade group illustrates how students used the problem solving plan after one instructional session. Sam's solution shows that he did not remember to make a drawing (TIP) or differentiate well among facts, questions, ideas, and the answer.

> **Acre Sale** The owner of a tract of land containing 10,000 acres sold 3975 acres to Mr. Jones and sold 450 fewer acres than this to Mr. Smith. How much land did the owner still have after these sales?

Sam's Solution	***Commentary***
Facts	
10,000 acres total	
3975 sold out of 10,000 to Mr. Jones	*Read 450 fewer acres given.*
450 less acres than 3975 to Mr. Smith	*Put subtraction under ideas.*
Subtract 10,000 – 3975 – 450.	
How much?	
Question	
subtract	*These are ideas. A question*
Add up 2, subtract from 10,000	*asks for what you want to find out.*
Ideas	
10000 – 3975 = what – 450 = answer	
3975 + 450 = what – 10,000 = answer	

Check

10, 000	3,975	10, 000	*These are ideas.*
– 3,975	+ 450	4,425	*Check the facts.*
6,025	4,425	5,575 acres	*Did you find 450*
– 450			*acres less than 3975?*
5,575 acres			

The next example shows the response of an adult student to a problem situation that was a novel one for her. A teacher of special education classes, she had used idea listing repeatedly in a university course and the Problem Solving Plan four or five times. Numbers given in parentheses represent a liberal scoring of different ideas.

Situation A farmer raised chickens, ducks, and steer. He had two-thirds as many chickens as ducks and four-fifths as many chickens as steer. How does the number of fowl compare with the number of steer?

Facts (3)
Farmer raises chickens, ducks, steer.
2/3 ducks = chickens
4/5 steer = chickens
Chickens and ducks are fowl.

Question (4)
How does the number of steer compare to the number of fowl?
How many steer are there?
How many chickens and ducks are there?
* Are there more fowl or more steer?
* How many more fowl or steer are there?
How many less fowl or steer are there?
Are there less fowl or less steer?

Ideas (10)
1) c = chicken d = ducks s = steer 2/3 d = c = 4/ 5 s

2)

d		c		s
2 x 2/3	=	4/3 x 5/4	=	1 8/12
3 x 2/3	=	2 x 5/4	=	2 2/4
9 x 2/3	=	6 x 5/4	=	7 2/4

3)

d	x 2/3 =	c	s	x 4/5 =	c
18		12	18		14 2 / 5
21		14	16		12 4 / 5
10		6 2 / 3	10		8
75		50	62.5		50

4) 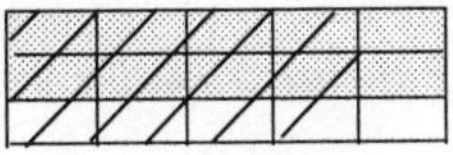

2/3 of d = c
4/5 of s = c
8/15 ?

5) 2/3 d = c   4/5 s = c
d = 3/2 c   s = 5/4 c

6)

c	x 3/2 =	d	c	x 5/4 =	s
10		15	10		12 2/4
8		12	8		10
6		9	6		7 2/4
80		120	80		100
40		60	40		50

7) There are more fowl than steer. Examples:
8 c + 12 d = 20 fowl and 10 steer.
80 c and 120 d = 200 fowl and 100 steer.
40 c and 60 d = 100 fowl and 50 steer.

Oh! There are twice as many fowl as steer! ANSWER. (2)

Total score: 19

DRAWING TO THINK DEEPER

Students decide how to make the general problem-solving plan and various representations of the question and solution help them find meaning. They must be willing to prolong thinking and trust that idea listing on paper will give them a chance to find connections among ideas and produce new ones.

Situation If 3/4 pound of candy sells for $2.00, what is the cost of a pound?

Some students write *3/4 x = $ 2*, where x is the cost of a pound, and solve the equation easily. Those who do not use algebra with ease, however, must rely upon original problem solving in which their ideas might not be organized well.

Although words and symbols are usually more efficient for routine thinking, drawings contribute to problem solving in ways that symbols do not. Students who remember to draw and are willing *to play around* with ideas use drawings to produce a new point of view.

Facts $ 2.00
3/4 of a pound

Question Find the cost of more than 3/4 pound.
* How much is one pound?
Find more than $2.00 for a pound.

Ideas $2 for 3 parts
$2 in three equal parts

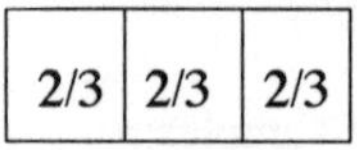

money

3 parts

Better get a whole pound

Four equal parts.

More than $2.00

Another part for 3/4
Last part is same price.

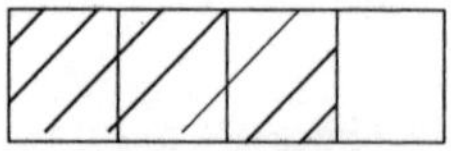

3 / 4 of a pound

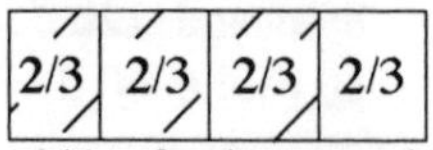

3/4 of 4/4 pound

2/3 + 2/3 + 2/3 + 2/3 = 8/3
$ 8/3 for a whole pound. ANS.

Some problem solvers are able to write *8/3 = 2 2/3* or *$ 2.67* immediately and understand why this notation is desirable. For others, *$ 8/3* is a logical answer. Some are content to estimate that the cost is somewhat larger than *$2.00* because a drawing with *2/3* labeled four times shows that the whole pound has *$ 2/3* more than *$ 2.00*. Others might make more drawings until idea production brings them to an equivalent for *8/3*.

Figure 1 Count the 8 parts. See 8/3 = 2 2/3.

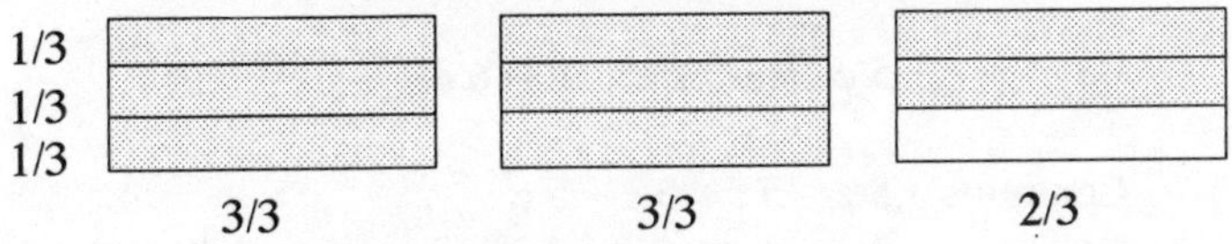

To many observers, profuse thinking is unnecessary or irrelevant. To the individuals who listed the ideas, however, the thinking process integrates prior activities and enables them to estimate and compare.

3/4 pound costs $ 2. Is the cost of a pound greater or less than $ 1.00 ? $ 2.00 ?

1/3 of 20 apples are ripe. Is the number greater or less than 20? Change 20 to another number. Greater than it?

20% off $ 4.50 Is it greater or less than $ 4.50? Is the cost greater or less than $ 4.00 ? $ 5.00 ?

Students can determine whether an estimate is preferred over an exact amount. In the candy situation, an estimate would usually be enough to help a shopper decide whether to make a purchase. To find a customer's bill, estimates are not acceptable. Drawings, imagined or written, often reveal approximations for quantities and serve to check computation.

In elementary school, there is scarcely a word problem for which students cannot find an answer to the question in a drawing. Time-saving is another advantage. Labeled drawings show facts and solutions that need not be written elsewhere in paragraphs and they often show relationships not evident directly through a text. For some problems that look novel at first, a drawing is enough. Without this help, a heavier burden falls on idea production. To find thirty cents worth of coins with the same number of nickels and pennies, for example, a drawing or a listing of the coins does the work (Figure 2 and 3).

Figure 2 Drawing

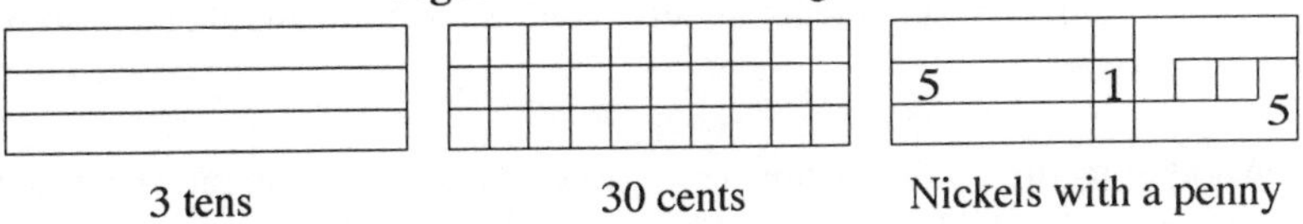

Figure 3 Idea Listing for a Word Problem

Begin:	5 ¢	and	1 ¢	It is 6 ¢.	
Continue:	5 ¢	5 ¢	5 ¢	5 ¢	
	1 ¢	1 ¢	1 ¢	1 ¢	Ten coins makes 30 ¢.

As in the drawing for *8/3 = 2 2/3*, a drawing for *3/4 + 1/3 creates* the sum. In this respect, number problems and word problems are the same. Students who manipulate numbers routinely to find a common denominator can learn how a few extra lines make it possible to count units. Fraction denominators suggest dimensions for rectangles.

Figure 4 Adding 3/4 to 1/3

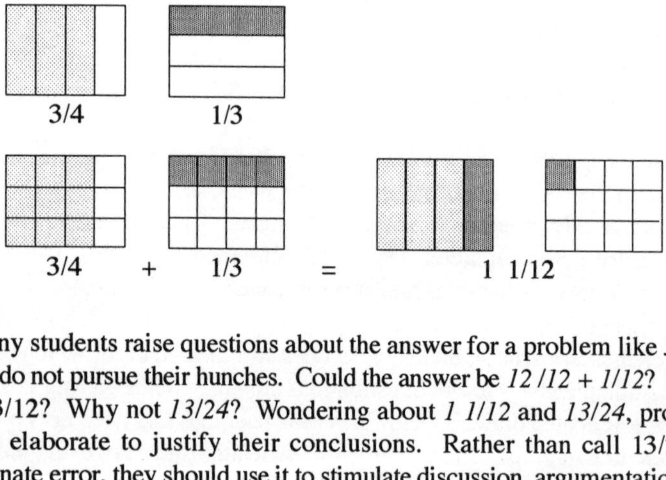

Many students raise questions about the answer for a problem like *3/4 + 1/3* but do not pursue their hunches. Could the answer be *12 /12 + 1/12*? Could it be 13/12? Why not *13/24*? Wondering about *1 1/12* and *13/24*, problem solvers elaborate to justify their conclusions. Rather than call 13/24 an unfortunate error, they should use it to stimulate discussion, argumentation and understanding.

Idea lists might also lead to a comparison of *ones*. If rectangles representing *24/24* and *12/12* are not the same size, then *12/12 + 1/12* is not *13/24*. Recall that *1 + 1 = 2* only if units are equivalent. Asking students for drawings, essays, or summary statements about a process encourages them to look beyond answers.

Another tactic is to show how concepts expressed in different words and symbols are alike. Making a 10 by 10 drawing, for example, students can show

how the same drawing illustrates many related concepts.

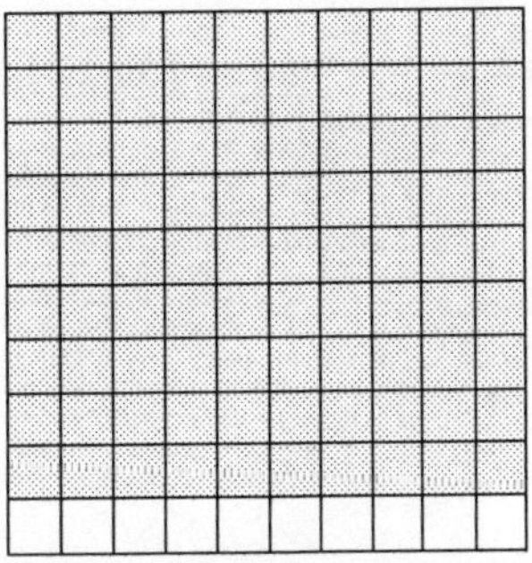

90 out of 100 boxes, 90/100, .90
90 pennies out of $1.00
90% of a rectangle

9 out of 10 rows, 1 out of 9 rows
9/10 and 1/10 of a rectangle
Ratio of 9 to 1 and 1 to 9

Length of 100 boxes: 100cm = 1m

The advantages of drawing and idea listing increase as problem solvers learn to adapt them to new situations. When an automatic response is inadequate, these strategies lead to an understanding of computation as well as problems in a context. To profit from this self-help, students must decide to think, expect to become confused, begin again, and sort out the best ideas. At the monitoring stage, they say, "I remembered to broaden the search instead of finishing as soon as possible." "I found that I could write things even though I first thought that I didn't know what to do." "I can help myself learn."

Students who free themselves from unnecessary constraints benefit from what they know and what they can figure out. Because routine thinking does not respond well to a dilemma, students should expect to develop strategies for thinking and allocate enough time to make them effective. For most students much mathematics becomes routine only when they have had an opportunity to use it often in a career or other adult experience. A problem-solving mode, on the other hand, requires a deliberate selection of strategies. Once this foundation for learning is established, it is likely to persist beyond classroom experiences to help individuals respond to a wide range of problem situations and new opportunities.

Notes

1. Giyoo Hatano, "Social and Motivational Bases for Mathematical Understanding," in *Children's Mathematics*, ed. Geoffrey B. Saxe and M. Gearhart (San Francisco: Jossey-Bass, 1988).

2. Thomas Peters, *Thriving on Chaos: Handbook for a Management Revolution* (New York: Alfred A. Knopf, 1987).

3. Howard Gardner, *The Unschooled Mind: How Children Think and How Schools Should Teach* (New York: Basic Books, 1991).

6 *Open Problems*

It is not unusual for students to treat word problems as a challenge well beyond their competence. The perception is that careful reading and sustained thinking offer no reward. Rules do not solve the problem and other people are not satisfied with their efforts. The advantage of open problems is that students take charge of mathematics by designing a context within their own mathematical knowledge and social experience. Invited to write a scenario about *one-half* and *four-fourths*, for example, many students speak about sharing a pizza. Confidence builds as students discover that they are able to produce more ideas than they anticipate and that their ideas make sense. With assurances that they can produce useful ideas about mathematics, many learn to challenge themselves.

A comparison shows the control that open situations offer problem solvers.

Closed Problems	**Open Problems**
Facts	***Facts***
• Sufficient data for a common solution.	• Little data leading to many possible responses.
• Students add facts to clarify given data and its context.	• Students add facts to establish a problem and place it in a context.
Question	***Question***
• Question given.	• No question given.
• Students restate the question to understand it better.	• Students state possible questions, select one, and restate it clearly. Questions often suggest more facts.

Ideas

- Students use free idea-listing and recall learned approaches. They identify a solution to the problem (question) selected.

Check

- For complex situations, students list the most pertinent facts, the question answered, and the answer.

The number and clarity of given facts regulate the degree to which students must add information. If a question is not included in the given scenario, problem solvers must design one. A complex question usually calls for the statement of more facts, including plausible estimations of quantities. It gives ample opportunity for students to construct interpretations and insights in the frameworks that they experience.[1] At the same time, students are relieved of conditions imposed on closed problems for only one acceptable answer and for precautions against assuming facts that are not stated.

ADDING FACTS TO MAKE A PROBLEM

A classroom experience illustrates how students handled a situation without much experience with open problems.

> **Situation** In the middle of the year, John moved to a school enrolling six times as many young men as women in the eighth grade. The enrollment for each class was about thirty.

Working in small groups, most students decided to find the number of men and women in each classroom. Several groups were unaware, or at least did not state, that this finding depended on an added fact, namely, that the ratio in each class was the same as the ratio in the school. One group that sensed the need for a class ratio abandoned the problem rather than add this fact. These students had not yet accepted the autonomy offered in open problems.

Another group concluded that *about 30* was exactly 28 students. Then each class had 4 women and 24 men. Some groups stated flatly that two students were "left over," and did not interpret this finding within the context of a real school. An imaginative group named the school Masculine Middle School and decided that two students were absent. They were male, the sex with the greater probability. No group provided a reason (fact) for the large male population.

Students agreed that they would be more aware of facts and assumptions if the scenario was part of their daily lives. Then they would get the information they needed. After some discussion, five categories of facts emerged: (a) data that makes a situation complex enough to be a dilemma; (b) information that makes a solution plausible, such as the ratio of students in a class equivalent to the ratio in the school; (c) information that makes a context familiar, such as names and dates; (d) axiomatic assumptions like the commutative property of multiplication; and (e) provisions for reliability, like accurate counting. In open problems, complexity and ambiguity usually require the most attention and establish the degree of challenge that problem solvers will tolerate.

CHANGING FACTS

Teachers who develop open problem solving encourage students to define missing facts arbitrarily and change facts that interfere with their ability to find solutions. A fourth grade student assumed that a raise in salary applied to the preceding year rather than the following one. This child was unaware that he changed salary conditions. Assuming a fact in ignorance or inserting it to make a solution possible takes priority over abandoning a problem without an attempt to solve it. A statement of these facts is constructive problem-solving behavior.

A college student knew how to find volume when three given dimensions had the same unit, but did not trust herself to convert inches to feet. Knowing the cost of cement for a sidewalk 64 feet by 4 feet by 4 inches, she listed the cost of a square foot among the given facts. She reasoned that the cost would be more than the cost of a cubic foot because the height of a normal sidewalk is a fraction less than *one*. Multiplying by a fraction like 1/2 decreases the cost. If the cost by volume was $60, then the cost of the same cement by area would be more than $60.

$$\begin{array}{lrcl} & \textbf{Volume} & & \textbf{Area} \\ \text{Because } h < 1, & l \times w \times h & < & l \times w \\ & 64 \times 4 \times h & < & 64 \times 4 \\ & \$\,60 & < & \text{cost by area} \end{array}$$

As a practical response in school, her thoughtful estimation of the cost of a square foot recognized the need for common dimensions and made problem solving possible. Estimating a fraction to represent the height of the sidewalk would reach the same goal. *What I don't know* leads to self-directed learning or to deferred on-the-job learning in the cement business. Self-directed students understand that they are the ones who make the decision.

STATING QUESTIONS

In ***SDM*** programs students evaluate their choices. The questions they select for open situations usually point directly to the intellectual risk students are willing to take. *Did I select an easy question or a challenge? Is it mathematical? Is it useful or interesting?* Usually, the more diverse the given facts are, the more creative the question. Students who initially interpret forty elephants and a computer as unrelated data force a relationship through a question. Problems emerge when they add facts, perhaps *.1 second for the computer to record the weight of each elephant* and *an average weight of two tons.* Practice in estimation, research on elephants, and knowledge of computers are by-products of the learning process.

For imaginative students, a drawing suggests questions at many different levels.

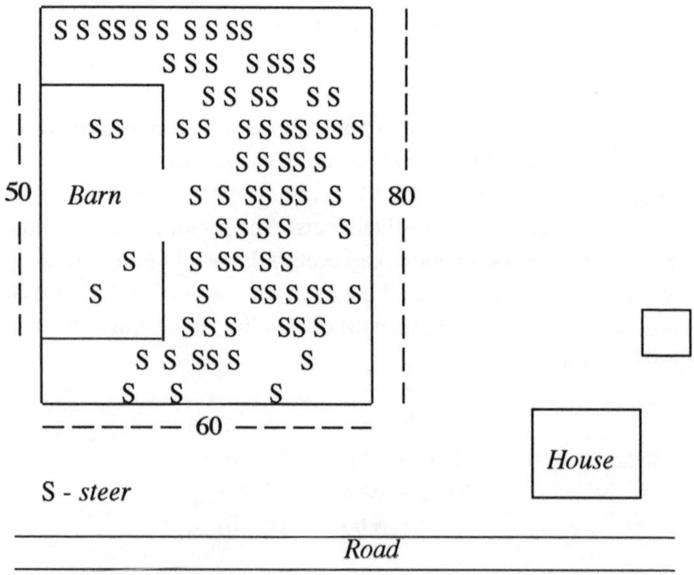

A college group added facts freely to support a variety of questions.

1. Is there enough space for steer to stand comfortably in the yard?
2. How many steer fit in the barn?
3. Would the south or east side require more fence to enlarge the corral?

4. How many minutes are saved by employing a larger branding crew?
5. How much new fence must be purchased for added steer
6. What percent of the steer fit in the barn?
7. How long did it take the steer to move into the road when the fire started?
8. How many steer per water trough?
9. How many bales of hay do the steer consume in a year?

Sarah was one problem solver who limited questions to the number of steer in a given area. She did not consider other relationships.

Facts

76 steer, 4 steer in the barn
Barnyard 4800 sq.yd.
Barn 50 x 22 1/2 (1125 sq.yd.)
House 110 yds. away
20 yds. from baryard to road
House is 11 yds. from road.
The building next to the house is a garage.
There is open area that could be made into more barnyard.
A lot more steer can fit into the barn.

Questions

How many steer fit inside the barn?
How many steer fit in the barnyard?
How many steer fit in one square yard?
What is the greatest number of steer the farmer can have in the yard?
* What is the greatest number of steer that will fit inside the barn?
Will all the steer fit inside the barn?
How many more steer in the barnyard than in the barn?

Ideas

9 steer fit vertically and 8 steer horizontally.
Keep adding more steer until no more fit.
Approximately 70 to 75 steer fit in the barn.
Find the area of 1 sq.yd. of steer and divide it into 1125 sq.yds.

8 x 9 = 72 steer will fit.

S S S S S S S S S
S
S
S
S
S
S

Sarah could have selected a more complex problem than one suggested by a simple observation of the drawing. A willingness to think further might have led to possible constraints placed on the situation by the number of animals and the dimensions of both an average steer and the barn. *Will 75 steer fit in that barn?* Moving back and forth among facts, questions and ideas helps problem solvers find an interesting question. For Sarah, this challenge would come later.

STUDENT SCENARIOS

The steer problem illustrates how a drawing suggested facts and questions. Other strategies used to design a mathematical situation are group brainstorming, directions to systematize the process, and a simple request for a practical context dealing with quantity. Prior experience with idea production and self-direction influence the effect of each strategy.

Children in the fourth grade began with only two familiar numbers, 10 and 100. One of them mumbled, "10 years. 100 inches." Other students said, "What do we know about this?" "What do we want to find out?" "In ten years I will be taller." The interchange continued with students contributing facts, questions, and ideas. The outcome was a thoughtful answer to the question, "How tall will I be in ten years?"

Without direction, this group had designed a plan for finding a problem.

1. Choose something with numbers.
2. Say something about it. Anything.
3. Write facts, question, ideas.
4. Ask, "Did I learn something new?"

In a fifth grade class, students living in a low-economic neighborhood worked on an open situation about bicycles. Directions more specific than the age-height plan helped them formulate a problem situation.

Directions

1. Write a ***Fact*** about bicycles. Put a number in the ***Fact***.
   Write another ***Fact*** with a number.

2. Think of a ***Question*** about the ***Facts.*** This is something that you do not know right away. It takes more than one mathematical sentence (equation) to answer it. Write the ***Question.***

3. Add more ***Facts*** to clarify something.

4. Rewrite the ***Question*** clearly. Answer it.

A group of four students worked together to create a situation.

> I have 7 ten-speed bicycles and 5 banana seats.
> There are 3 training wheels. How many do I have?

A student in the class quickly answered "Fifteen." After a brief discussion about the meaning of assumptions, the teacher drew a bicycle on the chalkboard. The drawing showed a ten-speed bicycle with training wheels and a banana seat. Then the teacher asked, "What is the greatest number of bicycles? The least number?" One student declared that ***Facts*** would have to include, "Ten-speed bikes do not come with banana seats." A careful analysis ensued as individuals described their interpretations of the situation. One girl remarked that these problems were "tricky." The inference was that students should think differently than they do for regular textbook problems.

Working in small groups, students in another fifth grade class designed problems with fractions to be solved by the entire class. This was open problem solving in which the use of a fraction was the only requirement. Using small numbers for the most part, students attempted to write stylistic, textbook expressions. Mathematical notation and the requirement to use fractions probably interfered with the thinking of these struggling writers.

- Walter went to the store with 3/4 of two dollars to buy some bread. The bread cost him 2/4 of a dollar. How much money did he have left?

- An item costs $14 and you pay 1/7 every day. How long would it take for you to pay for it all?

For the latter problem, the teacher asked students to substitute numbers for $14. Seeing that different costs did not affect the answer led to a generalization about 1/7 and a total, 7/7. This was a learning experience that might not have occurred without such a *strange* problem situation.

Other problems posed by the group raised more concerns about mathematical understanding and student communication.

- Gina went to the football park with 1/2 of 4 dollars. It cost 1/2 of 4 dollars to get into the ball park. How much did she have left?

- Lloyd ate 1/2 of a pizza and Aurelia ate 1/2 of the pizza, too. How much did they eat in all?

The teacher was amazed that individuals in the class produced different answers to each question. He wanted to know how students interpreted the data and whether they read with an intent to understand. Open problem solving suggested that error and interpretation carried nuances that are usually neglected when students work on closed problems. Some teachers claim that open situations should be introduced in school before closed problems. Then they would be able to study the natural thinking of students, and students could develop confidence in their own ideas before they are constrained by closed situations.

A college student intrigued with the notion of designing her own word problem described a situation about dustballs on the floor. After some exploration, she abandoned her original focus and settled for a time-area-number relationship. Like many descriptions of challenges in life, the tale was neither organized well nor complete.

> Mr. Dustball rolled from one end of the house to the other. The house was 62.8 feet long by 32.8 feet wide overall. Along the way Mr. Dustball met 20 dustball friends who traveled with him. They started the journey on a Monday.
>
> Mr. Dustball carried 7 dusty snacks. Each friend he met had 3 dusty snacks each. (Mr. Dustball always planned for emergencies.)

How big was the house?
How many days would the snacks last?
How many days would it take to travel from one end of the house to the other?

—————— 62.8 ft. ———————

32.8 ft.	Mr. Dustball's journey - - - - - - - - - - - OOOOO Balls OOOOO Friends, friends, . . .

In a subsequent discussion, the student remarked that thinking about quantity in this way was a liberating experience because she was in charge of her own thinking. Surplus data made the situation real for her and extended the idea-producing process. More thinking about the dustball problem would enable her to connect information and limit the problem to a selected area of interest. The goal is to make the story coherent and to show why a result answers a designated question. Open problem solving is an opportunity to develop this kind of synthesis.

WRITING STYLE

Writing style affects the design of a scenario and ways to find solutions. Daniel, about twenty-six years old, directed plays in the city and acted in them. After organizing some given facts in a problem, he used sentences in most of the remaining work. His choice to present the same facts in different formats showed an effort to explore the situation thoroughly, as directed, and a natural inclination to use language freely.

Situation Latoya wants a job. One job offers \$2 for every \$9 of goods that she sells. If she works for \$3 on another job, she gets to keep \$1 every time she earns \$3.

Facts		Latoya Sells	Latoya Earns
	A	\$9.00	\$2.00
	B	\$3.00	\$1.00

Both jobs are for the same amount of hours per week.
Latoya is offered both jobs, but can only work at one of them.

At Job A Latoya will sell $9.00 worth of goods in three hours.
At Job B she will sell $3.00 in about 30 minutes.

Question
* If Latoya wants to take the job for which she will earn the most money, which one will it be?
Which job will give Latoya the most money?

Ideas
Latoya will earn different amounts at each job.
In three hours at Job A Latoya will earn $2.00 because she earns $2.00 for selling $9.00 worth of goods.

In the same amount of time at Job B Latoya will earn $ 6.00 because she earns $1.00 for every 30 minutes of selling $ 3.00 worth of goods.

	Job A 3 Hours	Job B 1 Hour
Sold	$ 9.00	$ 3.00
Earn	$ 2.00	$ 1.00

Latoya earns the same amount of money at both jobs. She earns $ 2.00 at Job A after 3 hours, $ 2.00 at Job B after 1 hour.

After 3 hours at Job B, she will earn $ 2.00 x 3 = $ 6.00.
After 9 hours at Job B, she will earn $ 2 + $ 2 + $ 2 = $ 6.

Job B will earn $18.00 in 3 hours. This is 3 times the amount ($ 6.00) Latoya could have earned at Job A in the same period of time.

It takes six more hours at Job A to earn the same amount of money Job B will earn.
* Latoya will earn 3 times as much money at Job B than at A.

With the exception of the charts, Daniel's presentation is a contrast to the analytic style used by mathematicians and the more abbreviated style used by most students. The extensive writing that Daniel used probably stimulated idea production for him and repetition helped him focus on the facts. He used sentences to explore what he knew and to prompt new connections. In spite of the efforts of mathematics instructors to impose a more direct approach to problem solving, Daniel's verbal style is likely to persist and be more effective than other styles.

The most common response to the data from other students was to use equivalent time periods for Jobs A and B. The most efficient problem solvers answered the question about the better paying job by observing that 1/3 is greater than 2/9. Although Daniel had worked with ratios in more than one course, he did not transfer fraction notation to this situation. He did make the situation more complex, however, by introducing a 3-hour rate for Job A and 30 minutes in Job B.

Some students said that it would be easier for them to work in a situation that had more realistic facts. They would understand the scenario better if they knew what the jobs were and whether the earnings were commissions. In contrast, other students maintained that the more factual information stated in a problem by someone else the more concepts they must understand. They thought it was better for students to produce facts. Open problems prepare them to deal with many life situations in which knowledge and information are incomplete.

NOVEL SITUATIONS

Students who work on a situation foreign to their experience depend on intuition and a deliberate search for new ideas. In novel situations thinking swings in many directions in an effort to break through ambiguity. The generation of different categories of ideas, frequent readings, and restatements of the question support the process. Practice with unconventional activities makes students conscious of these strategies.

1. List five numbers with a median of 7. Make three more lists.

2. Draw a triangle with sides 3, 4 and 8. Explain what you did and what you found out. (No such triangle possible.)

3. List many ideas and write two summary statements about $1\ 1/3 + 4/3 = 8/3$ or $2/3 - 4/6 = 0$.

To develop summary statements, students begin with whatever they can say. The first response is usually brief. Referring to 1 1/3 + 4/3 = 8/3, they might say, "1 1/3 and 4/3 are equal." "It is two of the 4/3s." "The left side of the equation is 8/3." When summary statements connect more than one initial idea in the same sentence, they usually convey a new thought.

Responses from two students illustrate how self-initiated directions can stimulate thinking in an unfamiliar situation.

Situation How is 4 x 32 like 4(y + 2)?

Student A

4 x 32 = 128 It is 4 and 4 x 8	32 x 4 128	4(y + 2) has *y* and + It is 4 times y.

4 x 32 and 4y + 8
Both have 4s and 8s.
4 x 4 x 8 and 4 x something + 4 x 2

Try a number in 4(y + 2).
y is 8 and 4 x 10 = 40
y is 20 and 4 x 22 = 88

How are 4 x 32 and 4(y + 2) alike?
What if 4 x 32 = 4(y + 2)?
128 = 4y + 8
120 = 4y and y = 30

* Why not 4 x 32 = 4(30 + 2) as 4(y + 2)?
I should have seen 32 as 30 + 2.
Then 4 x 32 = 4(y + 2) It's easy to see y = 30.

* y = 30 makes 4 x 32 and 4(y + 2) MORE alike.

Student B

Facts 4 x 32
4(y + 2)

Questions
How are they alike?
How do the expressions resemble each other?
What is the same about them?
What do they have in common?

* Use many interpretations to find new connections between them.
State connections between them.
What works like this?

Ideas
Talk about 4 x 32 first. Then 4(y + 2).
Look at wholes and parts.
32 four times is the whole thing.
y and 2 are parts.
Am I doing it?

Writing directions within the solution is a matter of choice. Problem solvers determine, for example, whether writing *Look at wholes and parts* focuses their efforts or distracts them from the task. The act of writing can fill a void or clutter the problem solving. Individuals need to pit the advantage of abundant self-talk and idea production against unwanted meanderings.

SOCIAL SITUATIONS

Older students narrow the gap perceived between school mathematics and everyday experience through job situations and handling money. Those who think that their lives are unique can learn to find unique solutions to problems. Attracted to novel experiences, young children enjoy comparing the length of their walking strides with the principal's. "It's fun to follow the principal down the hall to count his steps." Concerned more with visible outcomes than the most efficient way to find an answer, however, both groups tend to disassociate tests and homework from problem solving outside of school.

One way to make students aware of their problem solving habits is to have them use the ***Facts-Question-Idea*** plan to use opportunities well and to settle their differences throughout the day.[2] Tonya, an eight-year-old girl, solved a playground problem that she identified.

Facts

Only 3 balls
Tonya and 3 friends

Playing pitch and catch

Boys took balls for kickball.
Boys grab balls and run.

Girls tell teacher who believes that boys had it first at recess

No set rules for who gets balls

First come, first served on the playground.

Questions How might I . . .
get the ball back?
persuade the boys to stop taking our balls?
get more balls?
ask for the ball back?
convince the teacher to get our ball back for us?
convince the boys we can all play with the balls?
* keep the boys from taking the balls away from us?

Ideas
Threaten them.
Throw balls back and forth fast enough so boys cannot catch it.
Hold the ball close to our bodies with all girls bunched up so that the boys can't get it.
Run away from boys.
Play with the ball near our teacher so she can see when boys bother us.
* Make a ball schedule for recess.
Ask boys to play with us.
Tell the principal.

Criteria
It's smart. It won't get us into trouble. Teacher likes it.
Feelings not hurt. Everyone agrees to it. Works all year.

Best Choice
Make a ball schedule for recess. Ask my teacher about it and make a chart tomorrow. Schedule begins the next day.

Developing this kind of self-direction demonstrates the effectiveness of idea production and the utility of planning. With practice, the generation of alternatives to social dilemmas strengthens mathematical problem solving and life-long learning.

Open problem solving makes students aware of their own ideas and the ways in which they construct meaning. Teachers not only insist on idea production, they also accept student thinking and demonstrate that ill-structured efforts lead to better problem solving. The charge is to have students select and regulate problem solving strategies along with the personality dispositions that support them. When school practice does not interfere, students adapt strategies to opportunities in academic and social situations.

Notes

1. Jere Confrey, "Learning to Listen: A Student's Beliefs about Probability" in *Radical Constructivism in Mathematics Education,* ed. Ernst von Glaserfeld (Dordrecht, The Netherlands: Kleuwer, 1991), 111-138.

2. Sidney Parnes, *The Magic of Your Mind* (Buffalo: Creative Education Foundation, 1981).

7 *The Assessment of Student Thinking*

In the current system, students are conditioned to believe that tests reflect the most important learning in school. A call for new forms of assessment[1] reaches into an evaluation of a broad diversity of performance, so broad that it evaluates nearly every important outcome. Instead of confining outcomes to those that can be measured easily, the goal is to examine ways that students go about problem solving and other strategies that contribute to performance. Assessment is regarded as a means for nurturing the construction of new knowledge and for measuring its transfer to new contexts. An expanded role for assessment promises to lead to new forms of written tests and many other activities that help students and teachers evaluate enriched outcomes for instruction.

Many models for evaluation used outside of school differ sharply from the comparison of scores to a national norm and from test grades dependent upon a percentage of a total possible score. They often involve more than one person, for example, and provide a criteria for judging performance. Ratings based on a specified criteria influence future work.

Several models used in the broader society have counterparts in school.

- For scientific research, individuals publish their work and weigh the criticism from readers. Similarly, classroom participants use computer desktop publishing to evoke a response from a wider audience than one teacher. Publishing student writing and problem solving is a service to others as well as a means of assessment.

- A committee or review board uses designated criteria for awarding grants and prizes to researchers. In like manner, school committees rate science fair entries and select the best proposals for games in a fund-raising carnival. The Future Problem Solving Program has evaluators who score problem solutions from teams of students and suggest ways in which students can improve idea production.

- Portfolios evaluated for technique and aesthetic accomplishment represent potential for success in an art school or for support in gallery shows. Teachers also are beginning to ask students for portfolios that demonstrate problem solving done outside of school as well as mathematics completed in the classroom.

• In the business world, economic returns and customer satisfaction indicate performance. Restaurateurs monitor the types of food requested most frequently and listen carefully to customer evaluation. Students in the mathematics classroom are their own managers and customers, the ones who decide whether the knowledge and strategies they use are adequate. Other students assist in the evaluation as they participate in team activities.

INSTRUCTION/ASSESSMENT

Changes in assessment realign procedures for placing students in school programs and offering accountability to the pubic. Teachers want to know if students can learn to do the kind of thinking used in a particular program and if they have the corresponding dispositions for reaching its outcomes. Which strategies do students use when routine mathematics is inaccessible. Do students clarify and generate problem questions and supplement given data with probable conditions? Do students give themselves access to the knowledge they have and make new connections with it? Do they confront misconceptions? This information, confirmed by other sources, is worth more to students than the high reliability offered by traditional measures of performance.

Outside of school, legislators and employers also learn more from the level of thinking students use than from comparisons among groups and individuals. They want to know if students are prepared to function in the work force. They are interested in the effect of school on continued learning and accomplishment.

New forms of assessment bridge the gap between assessment and self-help strategies when they connect the assignment of points to criteria for performance. In the explanation of a mathematical concept, examples in more than one kind of representation (e.g. words and drawings) merit extra points. If relationships and patterns contribute to understanding, then the assignment of points reinforces student production of them. Awarding points for independent ideas and for a coherent synthesis of ideas brings students to sense-making they do not experience in closed problems that exclude choices and the clarification of conditions for transfer to real-life situations.

The primary basis for assessment is evidence of the construction of new knowledge. Believing that seldom is learning sequenced discretely and uniformly, teachers use the ideas that students produce to negotiate meaning. Each idea reveals something about the ways in which individuals think, and shifting the emphasis from factual recall to other forms of thinking provides an opportunity for interaction between assessment and instructional activity. Rather than assume that a test score signals closure for learning, teachers encourage students to ask how they might use knowledge to broaden their interests and to seize opportunities.

Because qualitative differences in students cannot be turned into quantitative differences by decree,[2] the best assessment of learning reaches into broad concerns: (a) How does this paper or laboratory experiment show that the student is a problem solver? (b) Did the student offer an explanation that was not required? (c) How did this knowledge influence the student's later work or the work of others? (d) Has the student offered other evidence that I have not requested?

For some learning objectives, a global assessment is more valid than points and cut-off scores. Because students constuct meaning uniquely, they use varied thinking strategies. These differences establish the need for self-checking and change the goal for a teacher's evaluation of their work. When students work together on a problem of interest, they each contribute their strengths. Like members of a research team, differences often contribute more to the total effort than similar strengths in all students. Self-evaluation and feedback from an interested community makes learning satisfying as well as serviceable to others.

When the major responsibility for providing evidence rests with students, individuals find valid ways to demonstrate learning that teachers have not offered. Whether or not diverse methods are a more reliable measure than the same instrument used with all students in a class is a judgment based on priorities. Does the use of knowledge in a real context represent learning better than paper-and-pencil measures? Is it better to count the number of course objectives met or to average test scores on common tests? Students raise similar questions about course grades for sections of the same course offered by teachers using different instructional activities and different written test items.

To encourage developmental learning, teachers ask students to retake a test or use another format to show that they understand concepts. The demonstration of learning is more important than the date on which the assessment is made. Using a second stage of assessment diminishes the negative effects of *retesting*. For some subject matter, the second stage of learning complex thinking is the test on a subsequent unit.

QUESTION FORMATS

As assessment becomes an integral part of daily classroom activity, it becomes a means for knowledge building. The construction of knowledge is a problem solving mode of learning in which dialogue among each student, the teacher, and peers is important. On learning activities and tests, students expect to make trials, generate unstated conditions, and use knowledge in new ways. It is an explorative style of learning and living.

Scoring procedures guide students toward the generation of ideas. *What am I expected to know?* becomes *What can I learn? Which ideas might be better?* Individuals explore challenges in novelty, complexity, and abstraction. As in assessment outside of school, they benefit from experience with question formats that demand many levels of self-direction for improving the quality of the ideas produced.

Formats

1. **Five ideas** List at least five ideas about _____.

   Score all ideas except repeated and erroneous responses.

2. **At least 30** List at least 30 ideas. A student lists 40.

   Score 30 ideas from among the 40 listed.

3. **Extensive lists** List as many ideas as you can about ____.

Score all ideas except repeated, erroneous and memorized categorical entries. In the latter case, 2 x 3 = 6 and 2 x 4 = 8 receive one point, not two. Opposite ideas (4 < 5 and 5 > 4) might be awarded one point also. Scorers make other judgments as well. (See Chapter 4.)

4. **One or more correct answer** Find the correct answer. Find more than one response or more than one way to represent an answer.

   Score ideas as in #3. (See Chapter 5.)

5. **Open problems** Add facts and a question.

   Score as in #3. Give extra points for the complexity of the question produced by the student. (See Chapter 6.)

6. **Known but novel** Answer a mathematical question that presents known concepts in an unusual context.

   Score points for trials, unusual responses and correct answer.

7. **Exceedingly difficult** Work on a problem that is complex or with concepts beyond the scope of the course.

   Score as in #6.

8. **Looking for more** Students give two answers for a question that inferred one answer.

   Score two points for two correct answers.

LISTS

For tests that consist of only one idea-listing question, the score is the number of unrepeated ideas, with the exception of errors. Unrepeated ideas are actually categories and subcategories of ideas. To offset an inappropriate percentage of the total score represented by an idea-listing question, "List at least 30 ideas," the maximum number of points could be 15 with one point for every two ideas listed. To develop spatial thinking, score an extra five points for a drawing. If an unusual approach with a strong mathematical basis influences the general character of an idea list, instructors weigh this factor carefully so that scoring for fluency does not penalize profound thinking. For the most part, however, students with the most outstanding ideas generate the longest lists.

The wide range of thinking levels used in mathematics calls for qualitative criteria much like the criteria that composition teachers use to score essays. Students enrolled in an introductory course on statistics had scores that ranged from 30 to 110. The quality of their lists illustrated differences between students functioning at minimal competence and those who used original and complex thinking. Although producing the most ideas usually required more advanced ideas, everyone had an opportunity to describe what they knew and what could do with their learning.

Grade	**Score**	**Quality**
D	30 - 49	Error, few categories
C	50 - 69	Ideas taken directly from a textbook
B	70 - 79	Many categories of textbook items with original elaboration in examples, ideas not taken directly from the textbook
A	80 - 110	Much elaboration, theory building, conclusions, uses, complex mathematics in many categories, . . .

Instructors base decisions about grades on the group median score, the range of scores, and subject matter expectations for the course. Many groups, elementary school through college, have one or two students who produce ideas in numbers that far exceed the median score. When the distribution of scores has a large variance, teachers adjust letter grades accordingly. The number of scores for each letter grade assignment need not be equivalent. Teachers judge whether the ideas listed on tests do indeed show the learning recognized for letter grades determined by point intervals.

Judgment about cut-off scores resembles holistic scoring in which teachers begin the grade assignment process by placing test papers in three sets: (a) long lists with drawings, (b) average length, and (c) brief lists. The next sorting process assigns papers to five sets from A to F. Some instructors do the sorting without counting ideas for a test score. For program placement and other decisions with long-range implications, the involvement of more than one evaluator is usually more acceptable than the judgment of a single teacher. For the assessment of performance on a unit of learning and problem solving, test authors can submit written critiques of their work.

Before students take their first idea-listing test, they practice during a class session on the topic chosen for the test or one of equal significance. The general rule is, "When in doubt, write it down." Points are awarded for ideas and errors are ignored, not penalized. Some teachers prefer to identify the main categories of responses as a study aid and as an indicator of scoring criteria. Categories develop a useful thinking strategy in students and bring similar ideas together for more efficient identification of repeated entries.

In the practice session, students score their own lists and compare categories with a peer. If they list ideas in groups, another group evaluates the scoring. The discussion centers on categories of ideas and original responses. A study of the instructor's assignment of points after a test helps students learn more about the thinking used to do mathematics.

CORRECT ANSWERS

Tammy, a graduate student, worked on a problem without access to a memorized set of rules to guide the process. In this framework, she showed that she could produce many approaches to the problem and persist under conditions of uncertainty.

Question What part of the square is shaded?

Respond with many ideas.

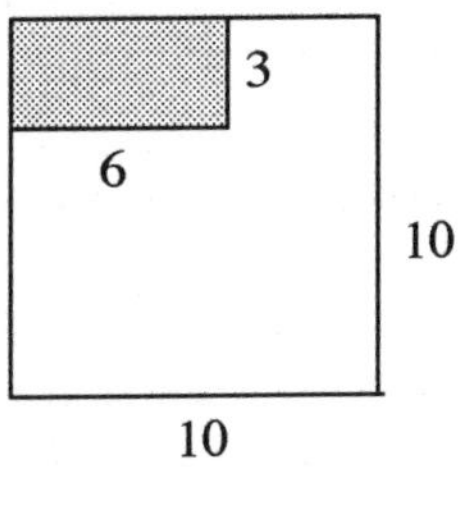

Response

1) Find, e.g., square inches.

10 x 10 = 100 sq.in.
3 x 6 = 18 sq.in.

$$10\overline{)100}^{\;10} \qquad 18\overline{)100.0}^{\;5.9}$$

90
10 0

5.9%

2) Find % of a square — 5 x 5 = 25 1 / 4

But how?

Obviously it is more than 5%.

$$25\overline{)100}^{\;4} \qquad 100\overline{)25.000}^{\;.2}$$

3) 18% ?

$$100\overline{)18.000}^{\;.180}$$

10 0
8 00

Score: 11 points

The teacher made some observations that would help the student solve many other problems. First, Tammy needs to determine whether she ever intended to understand square inches and percent. Second, the production of drawings would suggest new meaning for concepts and help her connect the concepts that she listed.

Other students working on this problem listed one or more final answers: 18/100, 18%, 9/50, 18 out of 100 boxes; 1/4, 1/5, 1/2, 6/100, a part.

OPEN PROBLEMS

Open problem solving is a constructive way to develop self-directed knowledge and to assess it. Students consider the conditions of occupations and other life experiences. At an early age, for example, they pay tax on their purchases and soon weigh the advantages and disadvantages of increasing taxes on different commodities and services. A written test question measures the mathematics demonstrated through student choices.

Test question Tax Problem

Would you rather pay (a) a 5% tax on radios, televisions, and washing machines or (b) a 2% tax on the purchase of a new car?

Describe a numerical example and give reasons or a purpose for your choice. Include many probable facts.

In the next step of the evaluation, groups of four students evaluate answers for the tax question written by two students not in the group. They decide which response is best and describe the criteria used to make the decision. Usefulness and depth or complexity are included in their critique. In the beginning, students might prefer to use the process anonymously with only the teacher knowing the authors of each response to the question. If the assessment contributes to a course grade, teachers assign the grade after the peer review.

One goal for the tax problem is to give students practice in the production of considerations that are not explicit. Students should reach beyond given data without instruction from someone else. Although in life situations no one says to base a decision on purpose, reasoning, and unstated conditions, problem solvers should generate them. As students mature, the paragraph in the test question about describing a numerical example and reasons should be unnecessary.

Another way to use peer review is to have students generate open (or closed) questions. In groups of four, students in grade five asked another group to solve the questions they wrote. (See Chapter 5.) Many groups were astonished that another group solved problems that differed from their intent. The activity is an assessment of team work among the authors and the clarity of their communication with others. Both the authors and those who solved the problems negotiate meaning and select the best problem for publication or distribution to yet another group of students. In a modification of the process, students write a problem that they solve themselves a month later. With this experience, individuals often recognize the need for clarity in their work.

KNOWN BUT NOVEL SITUATIONS

A group of college students responded to a test question about area in many different ways (Figure 1). The question had familiar vocabulary but not a familiar context. Responses indicated that students had to search for meaning. Although a few might have encountered a definition for *square inches* recently, most students used an application or school experience from a few years to many years ago. Some said they were conscious of creating a definition. Others might have related the situation to a carpet purchase in which they marked off square yards to determine the cost. The course in which they were enrolled did not include formal instruction on drawing square inches, but did offer extensive experience with graph paper to illustrate the meaning of arithmetic operations, e.g., 3 x 14.

In this case, the novelty of the question might have suggested a license for multiple interpretations of the question.

- A literal reading of *inch* requires that sides of squares be the same length as an inch on a standard ruler.

- Dimensions of rectangles have a product of 5.

- *Square inch* is figurative; squares do not have sides measuring an inch.

- Square inch is used in a square grid that ignores the requirement for five squares.

A preference for only one reading of the question and one approach to the answer, however, limited the thinking of these problem solvers. In the undergraduate group, most students had more time to work than they used. Those who chose to stop thinking about test questions usually assumed that thinking longer would not be helpful or that their responses were adequate. To have all students give the same response, the question needed directions that asked for a literal mathematical interpretation. If the intent was to measure many diverse interpretations from each student, test directions should ask for multiple responses.

The first column in Figure 1 shows the number of undergraduate students that gave each response. The second column gives responses from graduate students who were elementary and secondary school teachers. Neither group included mathematics majors.

Figure 1 Test answers

Question Draw 5 square inches, estimating the length of an inch.

N N **Responses**

10 15

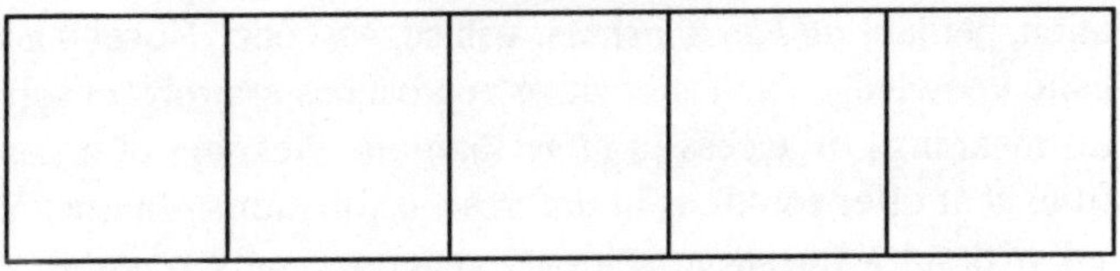

Square-inch boxes joined or with space between them.

9 5

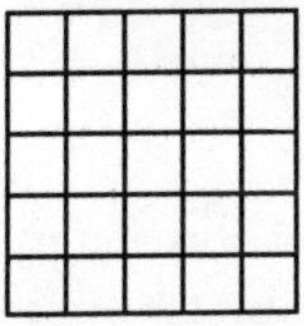

5 x 5 or 10 x 10 grids

4

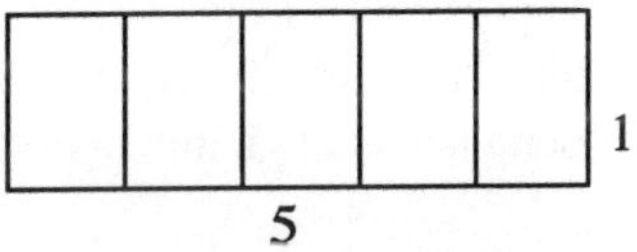

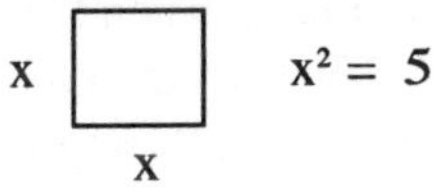

4 4

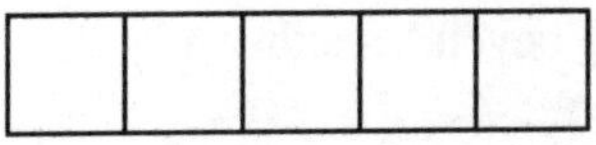

4 1

is about one inch.

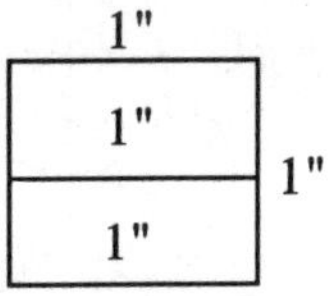

1

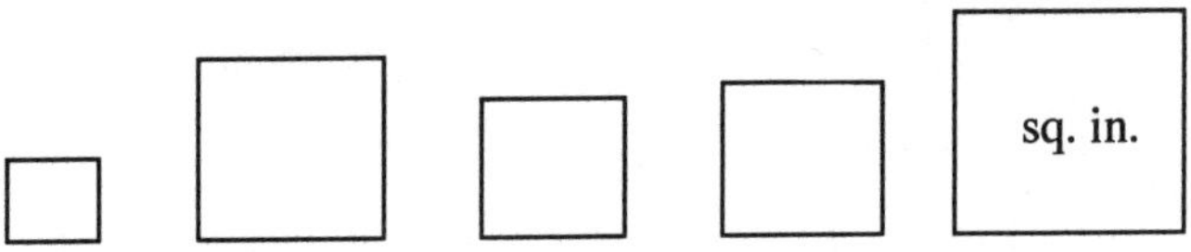

1

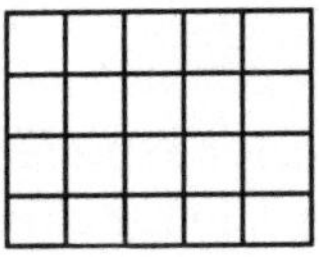

1 inch = 4 boxes
5 sq. in. = 20 boxes

A similar discussion deals with the statement of questions for word problems. Students need to recognize statements of the same question explained in different words and statements that describe different questions. In some situations, problem solvers produce more than one statement of a question deliberately. If *90 percent* interpreted as *90 out of 100* does not lead to a solution, perhaps *all but 10 percent* will suggest one. Novelty and lack of prerequisite knowledge call for creative approaches to problem solving, and nonliteral meanings or access to more than one meaning of a concept are possibilities that offer solutions in unforeseen situations. In open problems presented without a question, students generate many possible problems (different answers) before they select one for a solution. This format avoids expectations for the literal meaning inferred by traditional question formats.

DIFFICULT PROBLEMS

A third grade teacher in an inner city school gave scissors and paper disks to groups of four children. The problem was to imagine that the disks were five cookies for equal distribution to three children. Each group began by cutting the disks in half. Then the challenge was to cut a half into three equivalent sections. This was an opportunity to identify individuals who wanted to try more than one way to accomplish the task and had enough initiative to request more sets of disks for new trials.

Figure 2 Trials, thinking beyond, another way

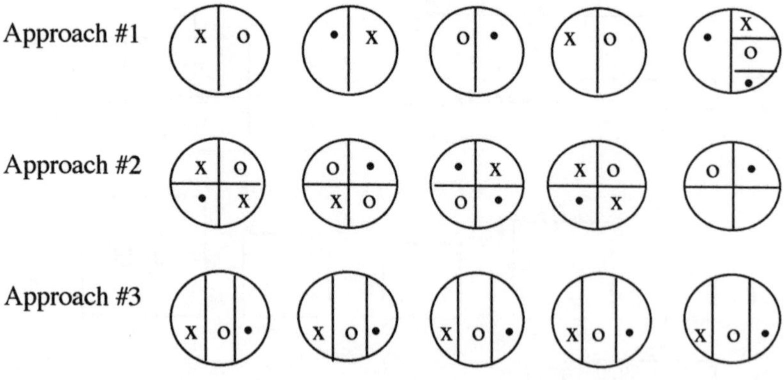

At the end of the activity, children drew their approaches to the problem on an overhead transparency. One child persisted at length in her efforts to cut quarters of a disk, but with only vertical and horizontal lines.

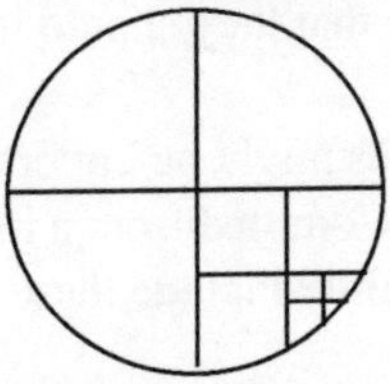

If students had interpreted a disk as a pizza or wheel shape, they might have said that it was easy to divide an entire cookie into thirds. Children in this group, however, confined their ideas to *cookie cuts* that did not produce parts that they deemed equivalent. Three parts of a cookie that did not *look equal* were not acceptable to them.

Assessment of mathematical thinking is enhanced by estimates accompanied by student explanations. The third approach in Figure 2 is reasonable when the outer sections are equivalent and the size of the middle section is adjusted to be the same area. Such a possibility has mathematical implications and practical merit.

LOOKING FOR MORE

A question about multiple groups suggested that students preparing to teach in elementary schools do not easily transform their thinking habits to those of mathematics teachers.

Figure 3 More than one answer

Name the mathematical operation illustrated.

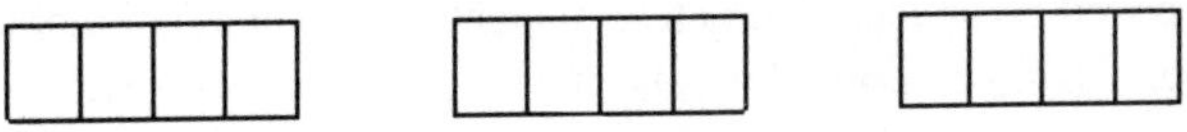

Addition, 4 + 4 + 4 = 12, was the most frequent response. Multiplication, 3 x 4 = 12, was the next most common answer. Only a few students listed both operations. The instructor of the course reasoned that if teachers did not think of both responses, they would not be likely to point out the relationship between addition and multiplication in word problems. A point was awarded for each response. In spite of the fact that students who listed only one response lost only one point out of a possible 65, they were distraught because the question did not state explicitly that they should list two operations.

On idea lists, one or two ideas might be important or fundamental enough to subtract points where they are omitted from a list. The logic is that a *good* list about a concept would certainly include these ideas.

Until this kind of thinking becomes an integral part of instruction, students might argue that they have been mislead into thinking that closure is preferred over a statement of possibilities or probable uses. Telling students to compare *81* with *100* is not the same as expecting them to create a life context or to produce quantitative benchmarks. "List ideas about *81*" asks for a reference to *50* or *100* only from students with number sense and the strategy to produce it. When classroom activities establish mathematics as a useful thought process in many frameworks, problem solvers produce ideas beyond initial answers. Perceptions of mathematics as independent thinking and exploration make the production of ideas beyond an answer a common practice for both children and prospective teachers.

DISPOSITIONS

Some students claim that tests are about their only experience with the generation of ideas under conditions of uncertainty. Others are overconfident: "I thought I knew what the chapter was about." If students allocate time for thinking and produce many options for answers, they are likely to decrease test anxiety and generate ideas much better than those who resort only to quick recall. A positive frame of mind prolongs thinking: "I know more than I think I do."

To score well on an idea-listing test, students have the job of making the test a challenge. Using advanced levels of self-direction, they write what they know and expect to find new insights during the test period. Their efforts resemble the initiative used to perform well in a career.

Idea-listing tests give students an opportunity to study for tests without the fear of unknown questions. As one college student put it, "I've had no experience filling a blank page, but at least I knew the question." Knowing a test topic days or weeks before an idea-listing test, students are likely to decide that they must understand principles and concepts. Understanding leads to connections produced during the test period and to much better mathematics than vocabulary and formulas alone.

To develop self-direction for both learning and its assessment, students complete a self-report instrument that stimulates reflection on thinking strategies and initiative. Monitoring helps them treat learning as self-directed problem solving regardless of the setting.

Self - Direction Inventory

Write the date on which you did the activities listed. Score one point for each instance. In two weeks, complete the inventory again. Compare your scores.

Date	Points	
______	_____	1. Reread directions carefully
______	_____	2. Asked for help
______	_____	3. Used a supplementary reference book
______	_____	4. Used a hand calculator
______	_____	5. Asked a question about a math concept
______	_____	6. Reread a paragraph for better understanding
______	_____	7. Listed 20 ideas to help me get started
______	_____	8. Made a drawing and labeled it
______	_____	9. Deliberately produced an unusual idea
______	_____	10. Looked for new connections
______	_____	11. Restated the question in my words
______	_____	12. Compared ideas with the question twice
______	_____	13. Decided to use more time and did so
______	_____	14. Estimated before or after computation
______	_____	15. Estimated instead of computation
______	_____	16. Produced alternative solutions
Total	_____	

Notes

1. See Gerald Kulm, ed., *Assessing Higher Thinking in Mathematics* (Washington, D.C.: American Association for the Advancement of Science, 1990).

2. Denis Newman, Peg Griffin and Michael Cole, *The Construction Zone: Working for Cognitive Change in Schools* (New York: Cambridge University Press, 1989).

8 A Changing World

To make the best use of opportunities, students either use creative problem solving or rely upon prescriptions. Rules applied routinely, however, do not address the many opportunities that involve unique conditions. Even principles do not answer questions in the here and now. For students who will enter situations unknown to them in school, problem solving requires idea production and new connections take precedence over following directions.

Because both knowledge and its applications undergo transformation in a rapidly changing world, groups from business, service, and other sectors have begun to ask for approaches to instruction that are based on inventive thinking. They clamor for access to alternatives, even new paradigms that rely upon initiative and original thinking as well as persistent learning.

Assuming that conditions will change, individuals prepare to improve what they do by monitoring outcomes, strategies, and new points of view. A cue from the business world is to offer support for trials and past failures so that people will experiment with new ways of doing things.[1] Journal writing maintains a continual flow of ideas and flexible thinking helps people select from, adapt to, and construct new environments.

Educators pursue change according to their perceptions of *what works* and their most current view of the world from the classroom window. Should schools transmit knowledge or ask students to construct it? Should they create a climate for self-directed problem solving? Should students prepare to initiate, plan, and unravel difficulties? Answers to these questions characterize learning environments and, for most students, have lasting effects.

To design the best educational experience for a group of students, long-range planners examine conditions in the workplace, the nature of a liberal education, and their beliefs about learning. The ***Self-Directed Mathematics Program*** asks further that they believe in the initiative students can use to help themselves learn and in the choices that students make, choices in the classroom that are as influential as the coaching that teachers do. Rather than assume that schools will offer a knowledge base adequate for novel situations, the program develops problem solving strategies and confidence that these strategies will develop mathematical power. This approach changes the expectations of students and the focus for instruction.

INDEPENDENT LEARNING

To take charge of their thinking, students produce ideas fluently and monitor their learning habits. Techniques for maximizing brain power help them recognize opportunities and use their own ideas to solve problems. An emphasis on the search for meaning and for principles to influence decision-making replaces rote methods that distract students from reasoning and invention.

When problem solving becomes a way of life in the classroom, it develops knowledge and new initiatives by offering ways to begin the process of working through ambiguity and ways to persist in the effort. Students who believe they are equipped to grapple with unforeseen situations are more likely to take advantage of changes in society and contemporary advances in knowledge. They are more likely to create situations from daily life and construct concepts than those who perceive that they have few options.

To examine the phenomenon, we can look at technology. Self-directed learners explore the capacity of machines and find new ways to use them. Computers give them a chance to discover strategies for doing things that they have not been taught and offer a compelling opportunity for individual invention. In homes, places of work, and every other sector, individuals figure out how to use technology to store and communicate all sorts of ideas and information. People who are highly dependent on others, however, find themselves disadvantaged.

Self-directed problem solving deals with challenges found in daily responsibilities and unexpected opportunities. The many employees who have lost jobs and looked for new ones, for example, depend on their own initiative to find a new way of life. They also identify the principles underlying their decisions. Unless they have a commitment to problem solving, both students and job-seekers fail to discover what they can learn and what they can do. *I will change my habits and find ways to use my skills* is much more promising than dependence on the recall of knowledge from a specialized situation examined in school.

CONTROL

The control of thinking is troublesome. Both teachers and employers struggle with a dichotomy between internal and external initiative when they look for the best locus of control for the most genuine learning and cooperative interaction among colleagues in school and at work. If teachers claim to know the best way to think, and the best way for all students, then they prescribe mathematical procedures in a precise manner. This position is based on a belief that algorithms are efficient and quick mathematics is better than prolonged thinking.

To increase individual connection-making and motivation, however, teachers must rely to a great extent on the self-direction of individuals. The most savory arrangement occurs when students adapt and design learning strategies as well as use routine mathematical procedures. "It's *me* thinking," they say. "I will find out what my brain can do." This intent differs greatly from attitudes that depend strongly on teacher directives for rote learning and well-defined criteria for performance. It also affirms individual accomplishment and the diversity that contributes uniquely to group efforts.

A long-term program is likely to find a very small percentage of students who must be controlled in nearly everything they do and a majority who profit greatly from much larger measures of self-direction. When students control their goals within a framework of finding problems and new insights, curiosity reaches into possibilities that develop new understanding and solutions to problems. As students learn to weigh the struggle of learning with the ideas they construct, self-satisfaction becomes as important to them as report cards from teachers.

Drawings are one of the most easily accessible and effective means for sensing personal control over learning because students can see the steps they use for developing an idea. The objective is for students to associate images with quantity and mathematical operations. Creating their own drawings, students see all the parts of a graphic and observe the effects of their own thinking on it. A drawing evolves as a gradual dramatization of an idea and suggests more ideas. When students recognize meaning of their own making, they not only find solutions to new problems, they also credit their own self-direction as an avenue for learning.

It is unfortunate that original drawings have been used only rarely in our schools, for graphics portray abundant data and make mathematics intelligible to experts and nonexperts alike. The synthesis and clarity of drawings develops and communicates the kinds of information that inundate the public and private worlds of individuals.

Idea listing without concurrent judgment is another means of internal control. Students who value their own ideas produce possibilities and alternatives that connect elements of prior knowledge and change faulty perceptions. They generate ideas freely to create drawings and enrich problem solving plans. An intent to consider many categories of ideas becomes as effective as checklists designed to increase flexibility. In a learning environment where students produce ideas freely, self-monitoring and evaluation contributes to the selection of strategies and problem solving dominates the use of procedural and conceptual knowledge created by someone else.

DISPOSITIONS

The search for ways to cut through an aversion to mathematics has begun to tap sources deeply rooted in individuals. Many programs use student choice and other forms of involvement to deal positively with emotions, beliefs, styles, motives, and identity as well as strategies and concepts.[2] Programs also make students aware of the expertise they use to learn mathematics outside of school without the benefit of formal teaching.

With access to a diverse curriculum, individuals stretch their abilities and interests. An opportunity to vary the pace and content of learning encourages them to explore the effect of self-regulated learning on achievement.[3] Students find out what they can do by themselves and with the help of peers. Another advantage of self-paced learning is that students need not suffer the constraints inherent in large groups, such as pressure from peers who wait for them to answer in class.

If students are to produce ideas and make decisions in school, they must change many of their learning habits. First, students must either create original responses or agree to prescribed substitutions for habitual ways of thinking. They not only become aware of their habits, but that there are alternatives and the choice is their own. Next, students must accept an educational paradigm grounded in mathematical understanding and independent problem solving. They decide whether to think for themselves about mathematics even while they benefit from external sources. The process is a preparation for life-long learning.

Habit	*Interpretation*	*New Response*
I don't know.	I cannot recall immediately.	I will say what I know.
I did it in my head.	I rely on recall alone.	I will monitor it.
We had that.	I do not want to understand.	I can understand more.
I have the answer.	I finish as soon as possible.	What else do I know?
I have the highest score.	The highest is the best.	Did I learn new ideas?

Journal entries written by a college student enrolled in a teacher education program illustrate the transformation.

> *First week of the semester* At this time, I am confused about what I will be doing in this class. I understand that I will be working at my own pace and choosing parts of my own assignments. I am not used to this type of math class. All through school we are assigned problems, have lectures, and do homework. I really never tried to understand math, just remember the concepts, rules, and algorithms. When I do math, I tend to scribble ideas all over, making trial-and-error guesses.
>
> *Eighth week of the semester* To my surprise I have found that this class

> and book have been very helpful. When I first started the course, I thought, "Hey, this isn't like any math class I've taken. How can I learn how to teach by making drawings?" Not only have I learned math concepts, how important understanding is, and new approaches to teaching, I've also learned how to learn by myself. It is so easy to fall into the pattern of asking questions and to have a teacher answer them. Rarely do you have to find out for yourself. But your class forces people to take charge of their learning. Not only do students need to ask questions, they have to work to find the answers also.

This was a student who was well on her way to self-direction. She was ready (a) to state objectives, e.g., find problems to solve; (b) to write plans for learning and for approaching novel situations; (c) to list mathematical alternatives and possible leads to solutions; and (d) to monitor goals and the effectiveness of strategies.

Knowing how to begin to solve problems in new contexts, students become thinkers who contribute to their own development and that of society. They are disposed to pursue new ideas and monitor comprehension through several strategies. Self-talk translates each strategy into a student's vocabulary.

- Skim topic sentences and pictures.
  "What is this all about?" (Out loud.)

- Relate information to experience.
  "What do I already know about this?"

- Predict possible outcomes.
  "I wonder if . . . "

- Read out loud.
  "I will concentrate on what I say and do."

- Make and check your predictions.
  "At first I thought that . . ., but now I think . . . because . . ."

- Ask questions to check your understanding.
  "Is there anything that doesn't make sense to me?"

- Summarize.
  "What it seems to be saying is . . ."

- Clarify.
  "How does this idea connect to that idea?"

- Resolve discrepancies.
  "I need to learn . . . and to find out how . . ."

- Seek the meaning of words from the context.
  "Which vocabulary words are new to me? What do they mean?"

- Check for challenge.
  "Was his problem a challenge for me?" "Too easy?"

- Produce abundant ideas.
  "I will list ideas without judgment for awhile."

- Use a plan.
  "I will list facts, questions, ideas and check for agreement."

PROJECTIONS

A disposition for self-direction and flexible thinking supports the changing nature of mathematical activity. Not only are there new kinds of situations and measurements, there is also a question of information that we know is missing. Even statistics are presented often with a caveat about the best information and methods available at the time. Researchers are aware that unknown data and circumstances influence us daily and that planning ahead calls for forecasting new patterns and new forms of quantification.

Dealing with the absence of data is not much different from clarifying muddled information that is disorganized or disconnected from it purpose. From this point of view, students learn by sorting through ambiguity and working with nearly all problems as if they are open problems. To learn new concepts and to solve problems means to make connections, often by adding likely circumstances and even inventing mathematical expressions to communicate events.

One approach is to establish an understanding that inferred data is that which has the highest probability extracted from statistical methods or expedient judgment. The effect on instruction carries a mandate for students to discuss, defend, and suppose what their work means in the light of extended data and multiple conditions. They do this alone and in teams. Correct answers are couched in a communication format that says: I consider probable and possible circumstances. The implication is that there is more to say, more ways to use the work, and more connections to make with other contexts and a later time.

With this approach, uncertainty gains respect and ambiguity becomes an expected part of mathematical problem solving. At the same time, new information and multiple connections to prior knowledge guard against contradictions. If ideas are missing, construct them. If information is inadequate and unavailable, determine the most likely scenario or prepare for stated alternatives.

Although the causes of unsatisfactory performance are diverse, one cause is surely that many students have little concern and few strategies for the development of understanding and the transfer of learning. Where the whole notion of learning does not include idea production, namely, independent thinking, students discount self-help and prior knowledge. The outcome is often an abandonment of learning as it is known in school.

At the point in their education when students want to know the importance of mathematics, they ask several questions. Which content is most important? Which applications occur most frequently or in certain careers? Should they connect ideas or will someone else do it? A broad view of mathematics shows that quantity comes in whole numbers, fractions, decimals, ratio and percent. Someone has the responsibility to select representations for quantity and patterns in symbols, words, drawings and graphs. Someone determines whether estimation or accuracy is most appropriate. With maturity, individuals conclude that *they* are the ones to make these decisions for specific situations.

Teachers nurture decision-making by asking for student-directed thinking that makes learning a matter of producing ideas to explore concepts and solve problems. Because, for most students, demands for quick, mental arithmetic interferes with problem solving, teachers face a dilemma. Should they emphasize immediate recall or provide for long-range thinking and learning? This is one of the most important questions educators will answer, and the nature of classroom goals and plans for assessment will convey their decisions.

As educators and employers throughout the country question methods that have been effective in the past, they ask for changes. With continued pressure for an emphasis on problem solving and reasoning, teachers will have to develop self-direction themselves and integrate it into the learning strategies they develop in students. Without change, students with excellent recall are rewarded academically but fail to prepare for challenges outside of school. An emphasis on idea production and self-monitoring, on the other hand, can compensate for a keen symbolic memory and prepare students to construct and use mathematics independently.

Perceiving the world as a dynamic place that offers new opportunities, educators prepare students to create knowledge and use it in a variety of unforeseen circumstances. Individuals produce ideas alone and in groups aware of the strategies they use and aware that their choices influence outcomes. In this scenario, curiosity responds to uncertainty and problem solving turns complexity and invention into opportunity. Guaranteed that future challenges will be unique, students learn best who learn self-directed problem solving during their first days in school and continue to develop it throughout life.

Notes

1. Tom Peters, *Thriving on Chaos: Handbook for Management Revolution* (New York: Alfred A. Knoph, 1987).

2. Geoffrey B. Saxe and M. Gearhart, eds., *Children's Mathematics* (San Francisco: Jossey-Bass, 1988).

3. Barry Zimmerman and Dale Schunk, eds., *Self-Regulated Learning and Academic Achievement: Theory, Research, and Practice* (New York: Springer-Verlag, 1989).

Index